Astronomy with a Small Telescope

James Muirden
Star Maps by Wil Tirion

Astronomy with a Small Telescope

Prentice-Hall, Inc.,
Englewood Cliffs, New Jersey 07632

© James Muirden 1985

First published 1985 by George Philip, 12–14 Long Acre, London WC2E 9LP

This edition published by Prentice-Hall, Inc., Englewood Cliffs, New Jersey 07632

Printed in Great Britain

This book is available at special discount when ordered in bulk quantities. Contact Prentice-Hall, Inc., General Publishing Division, Special Sales, Englewood Cliffs, N.J. 07632.

10 9 8 7 6 5 4 3 2 1

ISBN 0-131049941-2

Prentice-Hall International, Inc., *London*
Prentice-Hall of Australia Pty. Limited, *Sydney*
Prentice-Hall Canada Inc., *Toronto*
Prentice-Hall of India Private Limited, *New Delhi*
Prentice-Hall of Japan, Inc., *Tokyo*
Prentice-Hall of Southeast Asia Pte. Ltd., *Singapore*
Whitehall Books Limited, *Wellington, New Zealand*
Editora Prentice-Hall do Brasil Lta., *Rio de Janeiro*

Illustration Acknowledgments

The Moon photographs on pp. 52–3, 60–61, 66–7 and 72–3 are reproduced by kind permission of the Royal Astronomical Society.

Contents

Introduction

This book is designed to help people with absolutely no knowledge of astronomy whatsoever, and is intended as an introduction to the different celestial objects that may be observed with the naked eye, binoculars, or a modest astronomical telescope.

It includes night-sky maps showing the constellations visible month by month from two different latitudes on the Earth's surface; the northern one has been chosen for mid-European or North American sites, the southern one for observers in South Africa and Australia.

Chapter 1 gives some general information on observation and the selection of equipment. Chapters 2, 3 and 4 are about observing the solar-system objects – Sun, Moon, planets, meteors and comets – and Chapter 5 introduces the subject of stellar astronomy. The core of the book is Chapter 6, which gives information on the interesting objects to look for on a month-by-month basis through the year, all of which can be located by the detailed star maps. Chapter 7 is for those interested in astronomical photography, and discusses how you can use an ordinary camera to take photographs of celestial objects.

I am grateful to the well-known cartographer Wil Tirion for producing the beautiful maps that accompany the text, and to Lydia Greeves for being so patient and sympathetic an editor.

James Muirden

I
How to Look at the Sky

You will have bought this book because you want to learn more about the universe above your head and beneath your feet. It is a strange and frightening thought that we are surrounded by empty space, held on to the surface of our friendly globe by the invisible tug of gravity, and that this globe is, by astronomical standards, the tiniest speck of dust. If the Galaxy of stars, of which the Sun is a single member, were represented by a model the size of our Earth, our home planet would measure less than one thousandth of a millimetre across, and be utterly invisible even with a powerful microscope! But this should be a challenge, not a deterrent, for it is a wonderful thing that we have the intelligence and imagination to appreciate something so awesome.

There is plenty of 'popular' interest in things to do with space. Space launches and new discoveries about astronomical bodies often rate a place in newspapers and broadcasts. When an eclipse or a comet is due, there is plenty of warning. On the other hand, popular knowledge of astronomy is almost frighteningly low. Ask a friend why the Moon changes its shape, and the chances are that he or she will be unable to give a convincing explanation! Yet the fact that the Moon *does* pass through a cycle is something that everyone knows from childhood. It certainly does not follow that, because something is obvious and familiar, it is understood.

This book takes nothing for granted, except a willing audience, and it begins with the most basic and fundamental piece of equipment necessary for you to become a practical astronomer: your eyes.

What Can You See?

The human eye is an amazing piece of technology. It can focus on things as near as the printed page you are reading now, but also on stars and even galaxies across the gulf of space. It can see objects in brilliant sunlight or on a moonless night. It can sense colour.

It is, however, better at doing some tasks than at performing others. It is not specially designed for astronomy, for example. You may not find it very easy to see things through an astronomical telescope the first time that you look. This is because your eye is being required to do something different from run-of-the-mill searching and gazing. Many people are put off by this difficulty, and think that astronomy is a sham, or that they are half-blind! Persevere, and you will find that telescopic observation becomes more and more rewarding as the eye learns to see.

But there is no need to worry about telescopes when you are beginning to learn about astronomy. It is, of course, nice to have one, and you will not be happy for very long without one. But it is important to appreciate that plenty of astronomy can be learned without any optical instrument at all.

One of the finest relaxations is simply to stretch out under a clear summer sky and gaze up at the stars. With the aid of this book, you will soon know the names of dozens of stars and constellations, so that the sky will be a familiar place. You can then have the pleasure of showing the stars to friends, who may themselves have wondered what is 'up there', but have never taken the plunge (or the jump!) and tried to find out for themselves.

I have said that the eye is not designed particularly for astronomy. Yet it can still see plenty of stars, given the chance. The most important precaution to take, before hoping to see a fine night sky in all its glory, is to let the eye become *dark-adapted*. This is a phenomenon by which the *retina* – the screen on which the images that we see are thrown by the eye's lens or *cornea* – is made particularly sensitive. A fluid known as 'visual purple' bathes it when the light level drops very low, causing it to become enormously more responsive than in daylight conditions. Dark adaptation takes some minutes to achieve (in fact, experiments have shown that the process can go on for hours), so that it is useless expecting to step outside at midnight from a brightly-lit house and see thousands of stars, even if the sky is very transparent! Wait at least five minutes before even trying to see faint stars, and expect to find your vision improving further for some time after that.

The technique of *averted vision* can be useful when trying to view a faint object. This depends upon the fact that a zone about midway from the centre to the edge of the retina is the most sensitive to faint illuminations. Therefore, if you direct your gaze somewhat away from the object being sought, while concentrating your attention on that area, it may be brought into view. Averted vision requires practice, but a skilful observer may effectively double the aperture of his telescope when it comes to viewing faint stars and nebulosities!

People who wear glasses sometimes worry that astronomy is going to be difficult for them. This is not so. When staring up at the sky without any optical aid, a person with correctly-prescribed spectacles will see just as much as someone with excellent natural vision. It is not, admittedly, so easy to use binoculars or a telescope when wearing glasses as without them, but it is not usually necessary to wear them anyway, since the instrument can be focused to give a sharp image without spectacles being worn.

The only eye defect likely to prove annoying is *astigmatism*. An uncorrected astigmatic eye shows stars as short lines, or perhaps crosses, instead of as sharp dots or little discs. This defect cannot be eliminated by focusing, and stars seen through binoculars or a telescope may also show this distortion. In severe cases of astigmatism, it may prove necessary to wear glasses while observing; but many modern instruments allow for this, and permit you to observe comfortably in this manner.

Some Essential Equipment

Although some astronomical observation can be done during the day, most of it is done at night, and this means that you must be properly prepared for the dark and the cold.

There is no doubt at all that the first priority is defence against the cold. If you are cold, you are miserable; and if you are miserable you might as well pack up and go indoors. It may feel warm enough when you first go outside, but you will probably be remaining in a fairly stationary position throughout your watching, and a pleasantly mild evening can soon develop unsuspected depths of chill. Even if you decide against wearing any extra clothing at the outset of the watch, take some more outside with you, so that it is handy if and when you decide you need it. If you have to return to the brilliant lights of your house, you will lose that carefully-nurtured dark adaptation. (It is possible to preserve the adaptation of one eye by closing it, and using the other one to find your way around, but it is far better to make sure that a return to the house, once the observing session has begun, is never necessary.)

Apart from the obvious sweater or coat, think about the extremities of your body: head and feet. On a very cold night, a Balaclava helmet is excellent for keeping ears and face warm, but don't let it cover your mouth, or the reflected breath will tend to condense on telescope eyepiece or spectacles, dewing them up! The cold ground tends to soak warmth out of the feet, so two pairs of socks and thick boots are to be recommended for winter viewing.

A torch is essential, and it needs to be red and not too bright. Red light is best because, compared with other colours, it has the least effect upon the eyes' dark adaptation. Some observers use a bicycle rear light, but this is rather too bright, and tends to dazzle. You could experiment with a different bulb, or use a smaller battery if you don't mind altering the connections. Alternatively, and perhaps more satisfactorily, take an ordinary pocket torch, remove the shiny reflector from behind the bulb, and paint the bulb red using poster paint or perhaps a felt-tip pen. The effect of removing the reflector is to make the torch give a soft and even light, rather than the very intense but uneven beam that is normally produced.

You will need the torch to illuminate a star map, when you have one (more about this subject in Chapter 6), and also to illuminate any notes you may want to make. This brings us on to the important subject of *documentation*.

Recording Your Observations

In starting astronomy, you are embarking on a hobby that involves observing things. Some astronomical objects are easy to see; others are difficult because they are small or faint. Some are almost always present in the sky, while others are seen only fleetingly.

In the course of your study of the sky, you are bound to observe some rare and beautiful sights. Some will be precious only to you, while others may be scientifically important, and deserve to be reported. In both cases, I would recommend making notes about them. Making notes will help you to remember them afterwards, and may prove very important if you have seen something unusual. In addition, the act of note-making helps you as an observer, because you will find that you are having to think more deeply about what you are doing.

The world is full of disillusioned amateur astronomers! This is not because astronomy is difficult; quite the opposite. But there is so much to see in the sky that people often try to see everything, and are discouraged when they discover that this task is hopeless. If you spend the evening looking at clusters of stars, for example, and then pack the telescope away, you will remember only a confusion of sights when you wake up the next day, and will have no real sense of achievement. But if you keep notes about what you see, then you will have a most satisfying record to turn to later when you wish to refresh your memory, as well as a means of 'ticking off' the objects you have observed already, if you decide to make a special study of them.

So you need a notebook. I have found an ordinary large diary, with a whole page to each day, useful as a general reference to what has been done. But it may not have sufficient space to include a full night's work, and probably the best course is to buy a large stiff-covered notebook, quarto or A4 size, and write up the night's work in that, ruling off at the end of each session, or even starting each night on a separate page.

How you arrange your notes is entirely up to you: Figure 1 shows one way of doing it. What you must include, if the record is to have permanent value, are the following:

DATE: Year, month and day.

TIME: Using the 24-hour clock, and stating which time system is being used. In the British Isles this will be Universal Time (UT), which is at 0 hours at midnight. Do not use Summer Time, which is one hour ahead of Universal Time. If you are observing a series of objects, it is helpful to indicate the time at which each one was observed, since this could be important if, for example, a star suddenly flares up or fades down.

SKY CONDITIONS: Whether the sky is transparent or hazy; whether there is twilight or moonlight; whether the telescopic image is sharp and still, or 'boiling' and poorly

1 *A sample page from an observing book.*

TIME (UT)	OBSERVATIONS	NO.
	Sept. 11/12, 1976.	
16.50	Just managed to catch the Sun, 90mm aperture refractor, magnification x75. 3 groups seen.	4319
28.20	Another fine dawn sky. Venus brilliant, Jupiter and Mars high in Gemini. Dawn growing and clouding up fast.	
	Observed Jupiter with 12x40 binoculars. All four satellites seen:	
	· · O · ·	4320
	Sept. 15/16	
02.30 – 02.55	Jupiter observed with 12x40 binoculars:	
	· · O · Looked for its 4th moon Callisto, but not definitely seen.	4321
	Has α Orionis faded? I make it about 0.2 mag. fainter than Mars (mag. 1.1) and 0.3 mag brighter than α Tauri – but they are at different altitudes. (Naked eye.)	4322

defined. (The effect of the atmosphere on the definition of the image is known as *seeing* – see p. 25.)

TELESCOPIC EQUIPMENT (IF ANY): Details about the aperture, magnification, and type of instrument used. Of course, if you always use the same telescope you will not have to give full details each time, but can simply refer to it in some shorthand manner. But always refer to the magnification if it is sometimes changed: the magnifying power of a pair of binoculars is fixed, but any telescope worth the name will have a choice of magnifications available.

Observing Sites

It is natural to assume that the best place for an astronomer is deep in the country, far away from all artificial lights and industrial pollution. Certainly it is nice to observe from a site with a really dark sky, and it is worth while trying to experience such a sensation as often as possible, for the night sky seen in its 'natural' state is one of the most glorious sights available to mankind. But it is not, perhaps, helpful to start observing in such a sky, for the bright 'signpost' stars of the constellations are lost in the blaze of fainter ones, making them much harder to identify than when they are observed in the less pure skies of most urban and suburban sites.

Nor should you assume that it is possible to make 'better' observations in such a sky. Some astronomical work is not hindered by meteorological or industrial haze: observation of the Moon and the brighter planets can be carried out just as well from a town as from a rural site; and the Sun, of course, falls into this group. Some amateurs observe nothing else but these solar-system objects, and can lead a happy and satisfying existence even in the middle of a large town!

Even if you wish to observe stars, town lights may prove of little disadvantage. A sky may appear rather bright and hazy to the naked eye, but it darkens in a remarkable fashion when viewed through a telescope. This is an effect of magnification, so that telescopically you may be able to see stars from a town that are almost as faint as those detectable from a really good site, even though the naked-eye view is nothing like so impressive. A telescope, as well as being a star-brightening device, is also a sky-darkening device, and the sky-darkening effect helps bring dim stars into view. So do not despair if you want to observe faint stars from an urban garden!

The greatest problem for the town dweller arises when something faint and nebulous, such as a hazy gas-cloud in our Galaxy, or a distant galaxy far beyond the confines of our own, is being sought. These nebulosities can be elusive from a bright-sky site, since the sky-darkening effect of magnification, which brings stars into prominence, does not help much with nebulous objects. The brighter objects can certainly be seen quite well, even from a poor location, but there is no doubt that the country observer will enjoy the best views of faint objects. The same thing applies to those mysterious solar-system phantoms, the comets. But there is plenty to observe in the sky, even if faint nebulosities elude you!

Binoculars and Telescopes

Binoculars and telescopes are different forms of the same thing: a system of two lenses, one (the *object glass*) focusing an image of the object being observed in the air inside the telescope tube, and the other (the *eyepiece*) magnifying this image so that an enlarged view is projected on to the observer's retina.

Binoculars are a double telescope, permitting both eyes to be used simultaneously. They are relatively more comfortable to use than a telescope, since both eyes can easily be relaxed, and a set of prisms folds the light beam back on itself, making them very compact. However, since they are intended as a highly portable instrument, they are never very large, and their magnification is usually fairly low, as they are designed to be held in the hand. Since hands and arms are never very steady, the view will appear to tremble if the magnification is too great.

A telescope (and binoculars are included here) does two useful things:

1 It collects more light than the eye does, so that dimmer objects can be seen.
2 It magnifies objects, so that more detail can be seen in them.

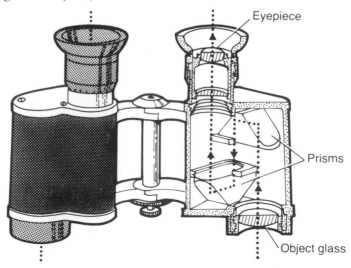

2 *Cutaway view of a pair of binoculars, showing their construction – a set of prisms folds the light beam back on itself.*

Eyepiece

Prisms

Object glass

Light-gathering Power

When you go outside at night, the pupil in your eye expands to its maximum value of about 8 mm across – in bright daylight it will shrink to about 2 mm or so. Therefore, when you look at a star, your retina is responding to an image formed by focusing the starlight falling on a circle only 8 mm across. Clearly, that star will appear brighter if the amount of light falling on a larger circle is collected and focused. This is what happens when a telescope is used, since the aperture of the object glass is much larger than that of the eye. Even a modest pair of binoculars will have a pair of object glasses some 30 mm across. Since the area of a circle is proportional to the square of its diameter, it follows that a 30-mm object glass will focus $30^2/8^2$ as much light as the 'night' eye. This ratio comes to about 15, so that a star viewed through 30-mm binoculars will appear 15 times as bright as when observed with the naked eye. An alternative way of looking at the situation is to say that these binoculars will reveal stars 15 times fainter than the faintest stars visible without them.

This is a huge difference, and means that we are looking at a much more impressive 'universe' even with very modest optical aid. It also explains why amateur astronomers like to obtain telescopes with the largest possible aperture. The smallest instruments that are sold as 'serious' astronomical telescopes have an aperture of 60 mm, and more ambitious observers usually have instruments in the 75- to 300-mm range, although some enthusiasts own even larger ones. (If you see an instrument described as a '100-mm telescope', the measurement refers to its aperture, not to its length!)

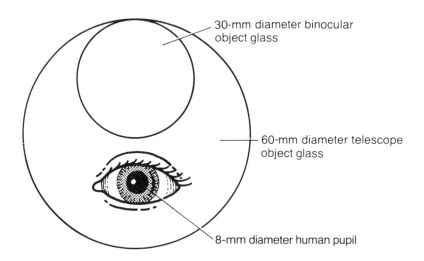

30-mm diameter binocular object glass

60-mm diameter telescope object glass

8-mm diameter human pupil

3 *The maximum pupillary opening compared with the object glasses of two typical optical instruments. It is obvious that even the binocular object glass will gather far more light than the unaided eye.*

x 20   x 75   x 200

4 The effect of magnification upon the field of view. This diagram shows the approximate size of the Moon, as seen with different powers.

Magnification

The magnifying power of a telescope tells you how much wider or higher an object looks when viewed through it. It is usually written as, for example, '×20', which means that the Moon would appear to have 20 times its naked-eye diameter.

Binoculars usually have a fixed magnification, although some variable-power or 'zoom' types are available. The smaller telescopes seen on the market may also have a zoom facility, but better-quality ones change their magnification by using different eyepieces. A typical 60-mm telescope may have magnifications or 'powers' of ×20, ×50 and ×100, but the value depends upon the manufacturer. Zoom eyepieces, though sounding impressive, tend to sacrifice quality of image for convenience, and their performance at the high-power end of their range is often poor.

It might seem that the highest magnification is automatically the best one to use, since the object will appear as large as possible. But this is not always the case. If you use a high magnification, only a very small amount of sky can be seen at one view. For example, a telescope charged with a magnification of about ×75 will probably be able to show the entire Full Moon at one view, with a little sky around the edge. The Moon's disc in the sky is about equal to an angle of half a degree, so that this is the telescope's *field of view** when used with an eyepiece of this power. With a magnification of ×200, the field of view will be about a quarter of a degree. On the other hand, a very low magnification such as ×20 can show two or three degrees of sky. If you want to look at a large area of sky, perhaps because there are beautiful arrangements of stars, or because it contains a large comet or nebulosity, then a low magnification will be better than a high one.

---

*Field of view is expressed in degrees or fractions of a degree (from the horizon to the overhead point or zenith is 90°).

Selecting a Telescope

The smallest useful astronomical instrument has an aperture of 60 mm. It is the type of telescope known as a *refractor*, since it uses a lens (the object glass) to focus an image of the object being observed. Some telescopes use a mirror instead of a lens, and these are known as *reflecting* telescopes, or *reflectors*, just as refracting telescopes are often referred to as refractors. Reflectors are popular with amateur astronomers who want apertures in the 100- to 300-mm range, since they cost less than refractors of the same aperture. With apertures of less than 100 mm, however, the refractor is universally used.

It is being assumed that you are using a 60-mm refractor to observe the objects in this book. There is nothing to stop you from using a larger instrument, and, obviously, the sights will be more impressive the bigger the aperture. But it is not necessarily a good idea to begin your observing career with a large instrument. A small telescope is more manageable, and is cheaper! There are plenty of superb sights available with an aperture of 60 mm. You will see an extraordinary amount of detail on the Moon; the phases of Venus; the markings on Jupiter; the rings of Saturn; more stars than you could count in a lifetime; numerous nebulae in our Galaxy, and other distant galaxies – plenty to occupy you for your first acquaintance with the sky.

Figure 5 shows a typical 60-mm refractor, mounted on a tripod. Looking first at the telescope itself, you will see that the object glass is protected by a

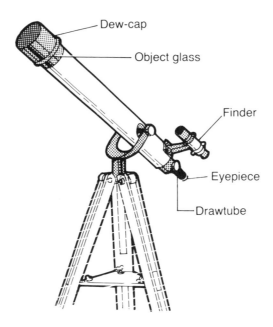

5 A small refracting telescope on an altazimuth mounting.

short projecting tube, known as a *dew-cap*. The purpose of this is to prevent dew forming on the surface of the lens when observing during damp conditions. In wet regions of the globe, such as the British Isles, a dew-cap is essential on many nights. Not all telescopes come equipped with one, but a cardboard tube or even a piece of rolled-up stiff paper, coloured black internally, makes an excellent substitute.

The eye end of the tube has a knob which can be turned to make a short narrow tube, into which the eyepiece fits, move slowly in and out of the main tube. This enables the image to be focused. Cheap telescopes often have rather stiff and jerky focusing knobs, which are a menace, because you cannot sharpen up the image without shaking the tube and making the view vibrate. Usually this is a sign of poor engineering, the manufacturer's method being to make the fitting very loose and to pack it with treacle-like grease. This prevents the tube from wobbling, but also makes it hard to move at all!

Usually a telescope has three eyepieces, although some have only two. These will screw or push into the focusing tube, and the magnification which they give should be marked somewhere on them. Eyepieces are often not treated with the care they deserve. They are small and easily dropped, but are as important as the much more prestigious object glass, since the best object glass will give poor results if used with an inferior eyepiece. It is advisable to keep eyepieces in a box, rather than in the coat pocket!

The other important feature of a telescope tube is the *finder*, a small telescope or sighting device which allows you to point the instrument at the required place in the sky. It is surprisingly difficult to bring even a bright object like the Moon into the field of view of an astronomical telescope, particularly if a moderate or high magnification is used. One difficulty follows from the astronomical telescope's habit of giving an upside-down and right-for-left image. There are historical reasons for this – old-fashioned lenses tended to absorb or reflect away an important proportion of light, and an erect image can be obtained only by using extra lenses. Since light is at a premium in much astronomical work, the telescope designer tended to favour a bright upside-down image rather than a dim correctly-orientated one. Nowadays, lenses are capable of transmitting practically all of the light passing into them, but the convention of the upside-down image persists, and most amateurs would be confused by observing the Moon or Jupiter the right way up!

A finder, then, is a convenient way of pointing the main telescope rapidly and surely towards a celestial object. A little telescope, with an aperture of about 30 mm and a magnification of about ×8, is ideal for the purpose. It should be adjustable, so that the centre of the field of view of the finder (often marked by cross-hairs) coincides with that of the main telescope.

Telescope Mountings

Most astronomical telescopes are mounted on *equatorial* stands, and the reason for this involves a little explanation. We observe the sky from a rotating platform – the Earth – which spins on its axis once a day. This axis can be imagined as an invisible line joining the north and south poles and extending into space. So it follows that the objects in the sky appear to rotate around this axis, once a day, although it is really the Earth which is turning.

You will have noticed, probably without thinking about it, that the Sun, Moon, planets and stars appear to rise somewhere in the eastern half of the sky and set in the western half. During the first half of their passage across the sky they are rising progressively higher in the sky, while during the second half they are descending towards the horizon again. They reach their greatest altitude when crossing the *meridian*, an imaginary line rising from the northern and southern points on the horizon, and passing through the overhead point, or *zenith*. Therefore, their paths are part of a curve, and to follow them with a telescope requires a tube that is adjustable both in *altitude* (up and down) and in *azimuth* (horizontally) in order to keep the object in view. A mounting of this type is known as an *altazimuth*, but this is rather awkward. An equatorial mounting makes life easier. This is designed so that the telescope only has to rotate around one axis in order to follow a celestial object across the sky.

This axis is known as the *polar axis*. It is adjusted, when you set your telescope up, to be accurately parallel with the Earth's axis. All you have to do, once you have found the object you want to look at, is to turn the telescope around the polar axis at a rate of one revolution per day. This is

6 Celestial objects follow a curved path in their passage across the sky.

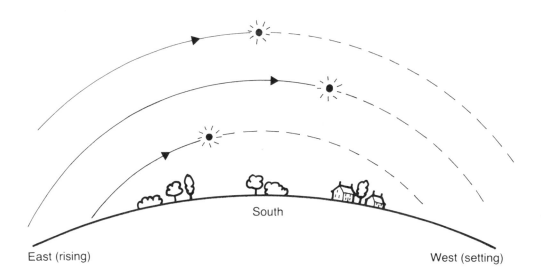

East (rising)　　　South　　　West (setting)

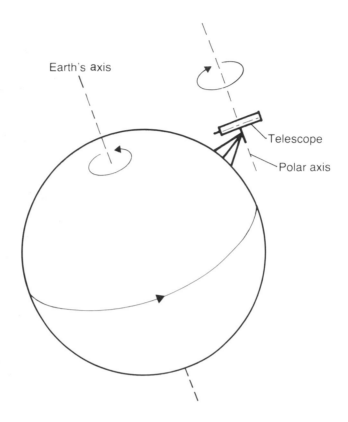

Earth's axis

Telescope

Polar axis

7 How an equatorial mounting works. By rotating the polar axis from east to west, in the opposite direction to the Earth's rotation, the telescope will remain pointing in the same direction in space.

often done by a small electric motor, although a hand drive, operating through a system of gears, can also be used. The polar axis is marked on Figure 7; Figure 12 shows the *declination axis*, which is at right angles to the polar axis and needs to be adjusted only when locating an object.

A good telescope deserves to be mounted on a steady and responsive mounting. Nothing is more irritating than a wobbly instrument, so don't waste time even considering one, since your astronomical career will be very short if you do. It is an unfortunate fact that some telescopes are designed and built by individuals who have no idea of the demands that will be made upon them. If you use this book to the full, you will find yourself spending many hours sitting in a dark garden, hunting up objects in different parts of the sky with the aid of star chart and finder; and sometimes it can take several minutes to locate a single object. If you do not have a steady mount, the result may well be a shaking blur on which it is practically impossible to rest your eye!

I have a theory, based on gloomy experiences with inadequate instruments, that the mounting of the telescope is more important than the optical quality. This goes heavily against the 'classical' order of priorities to be found in most books. The assumption in these is that the telescope giving the sharpest and clearest image is the one to choose, even if the mounting is not very good. However, I would suggest that the sharpest and clearest telescope in the world is useless, if the stand is so poor that the image cannot be properly grasped by the eye. Better a nice steady image, even if imperfect, than a shaking mystery!

You cannot judge a telescope by looking at its picture on a box. Insist on seeing it set up in the shop. Try turning the focusing knob, and see how much effort is needed. Does doing this make the tube shake, even to the eye? If it does, then observing will be a misery, and my advice is to leave the instrument alone. There is bound to be some slight movement – but if it is more than a millimetre or two as you move or *rack* the focusing tube in and out, then the signs are bad.

Try making the telescope turn around both axes. Usually this will be accomplished by twisting a knob, which engages a worm gear fitted to the axis. Is the motion sweet or stiff? Is it coarse or fine? If the telescope turns through a large arc with each twist of the knob, you will have difficulty in adjusting it finely on a star or planet when using a high magnification. The better the telescope, the finer and more secure this motion will be. Twist the knob both ways, to see how quickly the motion can be reversed – how much 'backlash' there is in the system. Again, cheap gears will have more lost motion than good ones.

Is the tripod tall enough? A drawback of refracting telescopes is that you need to look into the lower end of the tube, which can be awkward if the object being observed is at a high altitude. Your neck will be bent painfully back, and if the tripod stand is low it will be impossible to see anything without reclining in some ungraceful and uncomfortable attitude on the ground. The tripod should be tall enough to allow you to reach the eyepiece

8 Is the tripod tall enough – not too low, not too high?

even when the tube is pointing quite high in the sky. Many telescopes offer a small right-angled accessory known as a *zenith prism* or *star diagonal*, which enables the eyepiece to be fitted at right angles to the focusing tube. With one of these in place, you can look comfortably down rather than up, and this is a much more agreeable posture when observing objects at a great altitude. However, it is best not to use a diagonal if this can possibly be avoided, since it reverses the image top for bottom, and loses and scatters some light out of the image. Try to do without one if you can . . . which means buying a telescope with a reasonably tall tripod.

The tripod needs to be rigid. Put some force on the mounting, and see if the tripod twists or bends. If it does, it is not good enough – a telescope stand should be rock steady. No matter how good the mounting itself is, it is useless if it is sitting on the top of an unsteady tripod. Some tripods have a triangular tray set half-way down, which helps to stiffen the legs and bind them together. It may also be drilled with holes into which spare eyepieces can be dropped. If such a tray is supplied, it is important to test the telescope with it in position, since it is a structural member.

So you can tell quite a lot about a telescope just by looking at it. Rather like buying a house, it is easy enough to say if you *don't* like it, without even going inside. The 'outside view' eliminates a great many candidates. But it doesn't help to make a positive choice, because the optical quality of the telescope is extremely important, and this can be established only by trying it out, or by basing your choice on the recommendations of others who have used the same make.

Testing a Telescope

In fact, 'optical quality' is a minefield for the inexperienced and a godsend to the advertiser. In practice, you can get away with almost anything – and some manufacturers do! The reason is that there is no single, easily-understandable definition of what a good-quality instrument should do, and some dreadful telescopes masquerade alongside much better ones. Take this example from a recent test report on one 60-mm instrument, imported (like almost all the telescopes in this range) from Japan:

I'm sorry to say that the unit I tested performed miserably. A bright star image was misshapen, coloured, and astigmatic, using a power of × 150. A cheaper telescope of the same aperture, on a simple 'altazimuth' stand, performed much better optically. Under the stars the telescope performed very well . . . and I could just pick out markings on Mars with the power of × 100, later confirmed with a 15-cm telescope.

It is also the case that different samples of the same model can vary in quality, sometimes dramatically. Therefore, it is advisable to take steps to test a prospective purchase before buying it, or at least to arrange a water-tight refund should it not prove satisfactory.

The best person to make the test is an experienced amateur astronomer, who will know what to look for. Since amateurs are friendly folk, and will be happy to help anyone like yourself who is just embarking on the pastime they love, it makes sense to get in touch with a local society, if there is one, before you go ahead and purchase a telescope. Say that you want advice, and ask them to help you find someone who could help.

It may even happen that they know of a second-hand instrument, going cheap. But whatever happens, you will get advice and comradeship, and by joining the society and attending meetings you will quickly become involved, far more so than by confining your time to the reading of books.

If you have no local society, and must press ahead alone, the following advice may be helpful. (It is assumed that the stand (if an equatorial one) has not yet been set up very accurately; advice on how to do this will follow in the next section.)

1 Locate a bright star, not too high in the sky (at an altitude of 45° to 60°, say). Using the lowest-power eyepiece, bring it into the centre of the field of view. Now exchange the low-power eyepiece for one of moderate power, and again make sure that the star is at the centre of the view. Focus it as sharply as it will go. Is the image an intensely bright, very small spot of light? There should be no coloured fringes around it, and no rays of light spreading out from it.

2 Gently move the eyepiece a few millimetres inside the position of best focus. The image will expand into a small disc. This disc should have a fairly sharp edge, and be equally bright across its width. Now move the eyepiece outside the position of best focus by the same amount. The disc should appear similar to the previous one in size and light intensity. Note that the edge of the two expanded discs, even with a perfect object glass, will be slightly coloured – inside focus the fringe will be reddish or purple, and outside it will be a pea-green tint. However, when the star is focused sharply these tints practically cancel each other out, and only the faintest blue-violet halo will be seen. Any conspicuous colour at focus means that the object glass is useless.

If you want to see what a perfect star image should look like, take two small pieces of card and make a pin-prick in each one. Hold one at arm's length, towards a light, and view it through the other. You will see the so-called *Airy disc* and very faint concentric rings that are visible in an image seen through a first-class telescope.

3 Check for astigmatism by observing carefully the shape of the two expanded discs. They should be perfectly circular. If they are elliptical, then the object glass is astigmatic and useless: unless the fault lies in your own eye! If you normally wear glasses and have removed them, put them on and see if the trouble disappears. If in doubt, loosen the telescope tube

in its cradle and rotate it while looking at the star. Telescopic astigmatism rotates with the tube, so that the ellipses will appear to turn. If they remain in the same orientation, then the problem will lie in your eyes.

4 The star-test described above is the best 'optical' test. But if the Moon or a planet is in the sky, it is well worth looking at these objects too.

Charge the telescope with its highest magnification, and look at the Moon. First, examine its bright true edge, or *limb*. It should focus up as sharp as a razor, white and brilliant against the dark sky. Do not worry if there is a faint bluish halo, as there is almost certain to be – worry only if it is red or green. Now look where the shadows are long and the surface detail stands out in its rugged splendour. There should be brilliant, colourless contrast between the surface and the shadows; even with a small telescope, the image should look sharper and more contrasty than in any photograph.

Do not be distracted by a wobble or shimmer across the image. This is caused by atmospheric unsteadiness, or *bad seeing*. Some nights offer better seeing conditions than others. But, unless the seeing is ruinously bad and the image is a swirling mass of air currents, you should be able to detect a crispness behind the shimmer, which stands for a good telescope.

If you can locate one of the bright planets – Jupiter, or Saturn – take a look at them, too. Even with a magnification of $\times 50$ you will see a disc on Jupiter, and the rings of Saturn will be detectable also. The image may be shimmering, particularly if the planet is low in the sky, but it should be well defined. If it is woolly, the telescope is probably a poor one, and will already have failed the star test! Venus, after dark, is too bright and flaring to be much of a test, since it will probably look 'messy', even if the optical quality is excellent, due to excessive brilliance and inevitable low altitude.

5 During these tests, you will also be testing the mounting for steadiness. Does the image wobble violently as you turn the focusing knob? Is it easy to bring an object into the field of view? To make a proper test of an equatorial mounting, however, the stand needs to be set up accurately, and the next section describes how to do this.

Setting up an Equatorial Telescope

If you are buying a new instrument, it will very likely come with a book of instructions to help you set it up. Not all instructions are very clear, however, and if you are buying a second-hand telescope, you may have no information at all.

The point about an equatorial mounting is that it must be set up reasonably accurately, or else it will be worse than useless. If it is not accurate, you would be far better off with an altazimuth mounting, which requires no special setting up, since one axis is bound to be more or less upright, and the other one will be approximately horizontal. To set up an equatorial, however, you need to know the position of the Earth's axis, and the best way to find this out is to use the Sun and the stars to help you.

This may sound a major problem, but it is not all that difficult. Imagine that you are sited in the Earth's northern hemisphere, as shown in Figure 7. There are two operations related to adjusting the polar axis. First, it needs to be set up so that the upper end of the axis is pointing northwards, and the lower end is pointing southwards; this is achieved by turning the tripod round in azimuth. Then, the angle that the axis makes with the horizontal must be right; in fact, it must be equal to your latitude. If you are located in latitude 50°, the axis must make an angle of 50° with the ground. (In the southern hemisphere, the upper end of the polar axis must point southwards, and the lower end northwards; however, its altitude must still be equal to your latitude.)

9 BELOW AND RIGHT *The position of the celestial poles, with reference to nearby stars and constellations.*

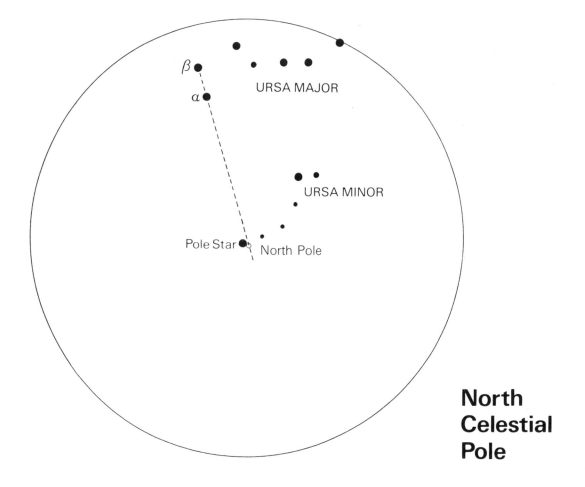

β

α

URSA MAJOR

URSA MINOR

Pole Star   North Pole

North Celestial Pole

Azimuth adjustment You can find out the north and south direction in several ways. If you can get hold of a surveyor's compass, use it to identify some distant landmark that can be used as a reference. Remember, however, to allow for the difference between magnetic north and true north – this will be indicated on a good survey map. If you know the instant at which the Sun is due south, you can cast a shadow using a vertical stick, and draw a north–south line on the ground. Unfortunately, it is not easy to obtain the Sun's true southing time without doing some calculations, since it varies throughout the year, and depends also upon your longitude; the Appendix gives you some information about this.

In some ways, it is easier to set the azimuth at night, provided the stars are visible. Using a star map (Fig. 9), you can locate the position of one of the celestial poles. The north celestial pole is easy to identify, since it lies within a degree of the bright Pole Star. In the southern hemisphere, the celestial pole has to be located with reference to nearby, rather faint stars, but it can

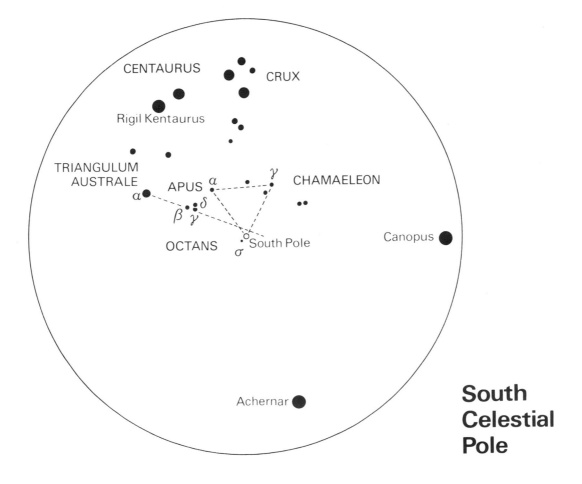

South Celestial Pole

be done. Get the polar axis aligned on this point by sighting along it towards whichever marks you have chosen.

Altitude adjustment In some ways this is the easier of the two, since it can be done using a spirit level and an adjustable protractor. Set the protractor to an angle equal to your latitude, lay one edge against the polar axis, and lay the spirit level along the other edge. Adjust the altitude until the bubble is central. It is not difficult to achieve an accuracy of about one degree by this method, and this is good enough for most observational purposes.

Fine adjustment You will soon discover if your mounting is not sufficiently well aligned, for a star or planet will rapidly drift out of the field of view if you try to follow it by turning only the polar axis! It is possible to compensate for this drift by making adjustments of the declination axis, but it is better to get the mounting right in the first place. Usually, azimuth errors are greater than altitude ones, because of the difficulty of sighting on a distant marker. So try twisting the stand slightly, and see if the tracking improves. If it gets worse, then you know to turn it in the opposite direction!

Once you have got the stand well adjusted, try to make the setting easily repeatable each time you go out for an observing session. Perhaps you can lock the head rigidly and permanently to the tripod, and make some marks in the ground where the legs have to go. Alternatively, you may decide to

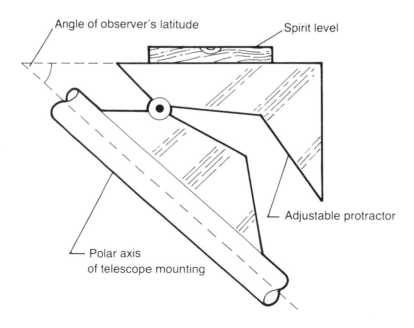

Angle of observer's latitude

Spirit level

Adjustable protractor

Polar axis
of telescope mounting

10 *In setting up an equatorial mounting, the polar axis can be set very close to the required altitude by using a spirit level and an adjustable protractor set to the observer's latitude.*

leave the tripod permanently in position, and relocate the head, using some marks, each time the telescope is set up. Anything which shortens the assembly time is to be welcomed, particularly if you live in a climate which is prone to cloud and in which starlight time is particularly precious!

Motor drives As well as checking the adjustment and rigidity of the mounting, you will also be able to test out the motor drive, if there is one. This usually consists of a small electric motor, which runs directly off the mains supply, and a word of warning is appropriate here: be extremely careful of current leaks. Telescopes are often used in damp conditions (some autumn nights, excellent for viewing, can leave exposed parts running with water), and the metal frame of the motor, or even the mounting itself, could become 'live'. It is important to ensure that all metal parts of the instrument are earthed, so that if there is a serious current leak a fuse will blow and shut off the supply.

Having checked this, and added an earth wire if necessary, switch on the motor and check how well the telescope follows the stars as they move across the sky. (As always, the stars are in fact stationary, and it is the Earth which turns, but we speak conventionally of the stars 'moving'.) A satisfactory drive will keep a celestial object near the centre of the field of view for at least half an hour – perhaps longer – although it may wander from side to side in the interval. This is irritating, but does not really matter unless you are attempting photography, when accurate tracking is very important. The better the drive, the smaller this wander becomes; it is caused by imperfections in the gears, and little can be done to improve it. If the star rapidly wanders from the field of view and vanishes altogether, the trouble is probably caused by slip in the clutch which links the polar axis to the drive wheel, and this will need to be tightened.

Some telescopes, particularly the more expensive ones, come equipped with a 'variable-frequency' control. A synchronous motor, which is the kind commonly used, keeps its steady speed because it is controlled by the 50 Hz (cycles per second) oscillation of the mains electricity supply (in the USA, the frequency is 60 Hz). By using a separate controller, which generates its own oscillating current, the speed of the motor can be changed at will. This is particularly useful if you are observing the Sun or the Moon, both of which appear to move across the sky slightly more slowly than do the stars, and will gradually drift out of the eastern edge of the field of view of a telescope set to rotate at star rate, or *sidereal* rate. Alternatively, you can compensate by switching the motor off for a few seconds every now and then, to let the object catch up with the telescope.

Other Types of Telescope

I have spent some time describing a typical refracting telescope, and have pointed out that an instrument with an object glass only 60 mm across is adequate for the observing described in this book. But it may be helpful to mention, briefly, the other types of telescopes that are also frequently used.

The reflecting telescope has already been mentioned, and Figure 11 shows the way the commonest variety, the *Newtonian*, works. A concave mirror reflects back the light from a distant object, forming an image of the object at some point in front of the mirror. It would be impossible to place the eyepiece here, since the observer's head would block the view, so a small flat mirror reflects the light through a right angle and out through a hole cut in the side of the tube. Therefore, when using a Newtonian, the observer looks in the side of the tube, side-on to the direction of the object. At first this may seem a very strange way of using a telescope, but it soon becomes not only normal, but extremely comfortable, since the eyepiece does not tilt at a refractor-type angle when observing an object high in the sky.

Another telescope design, perhaps less familiar than the Newtonian since it is only used for rather expensive instruments, is the *catadioptric*. The optical system in this telescope uses a very short tube (usually less than twice the width of the aperture), with a thin lens at the front and a concave mirror at the back. These telescope systems are attractive because of their extreme compactness, but do not actually give a better image than a first-class refractor or Newtonian reflector, despite what the advertisements may say!

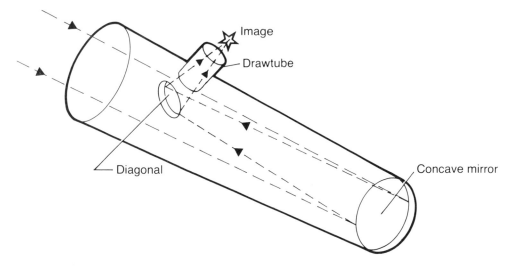

11 How the optical components of a Newtonian reflecting telescope are arranged. The concave mirror at the lower end of the tube reflects a cone of light back on to the inclined diagonal mirror. It is then reflected along the drawtube to form an image of the object at the side of the telescope tube.

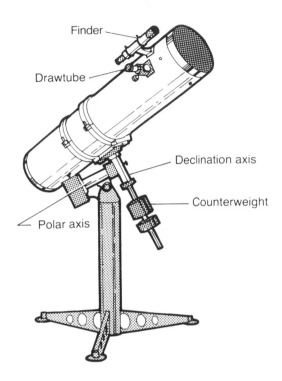

Finder

Drawtube

Declination axis

Counterweight

Polar axis

*12 A Newtonian
reflector on an
equatorial mounting.*

Celestial Motions

This section gives some very basic facts about the sky, and our view of it from this rotating planetary platform of ours.

The Celestial Sphere

The first thing to sort out is the way the stars and planets appear to be scattered around the Earth. They seem to form patterns on the surface of a huge sphere which rotates around the Earth. This rotation is an effect of our own planet's spin, which carries us around from west to east, and makes the sky appear to move from east to west. The 'huge sphere' effect is caused by the fact that all astronomical objects appear to be so far away that we have no sense of distance when we look at them – in a sense, then, they appear to be equally remote, and give the illusion of being fixed to the inside of a bowl. We call this the *celestial sphere*. It does not exist, except in people's imaginations, but it is a very useful thing to imagine!

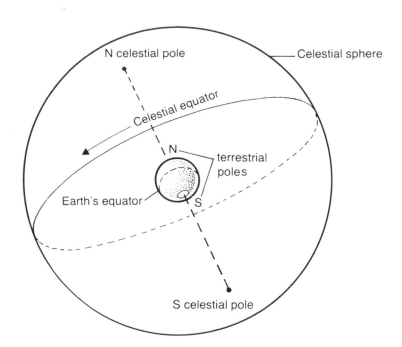

13 The celestial sphere, with the Earth at the centre, showing the celestial poles and the celestial equator.

There are three important markings on the celestial sphere. Imagine that the Earth's axis is extended infinitely far through space, so that it reaches the stars. The points where it would pass through the 'sphere' are known as the *celestial poles* – northern and southern, diametrically opposite each other. Imagine also that the plane of the Earth's equator is extended, like a flat sheet, to reach the celestial sphere. It would create a huge ring or circle around the sphere where it passed through it, and this is known as the *celestial equator*.

The celestial poles and equator are marked on all star atlases. The star patterns are also crossed by imaginary lines of celestial latitude and longitude, just as the Earth is divided up into similar divisions. Celestial latitude is known as *declination* or *Dec.*, and celestial longitude as *right ascension* or *RA*.

Northern and Southern Stars

An observer standing on the Earth's surface can, of course, see only half of the celestial sphere at any one time, since the other half is hidden beneath his feet. Figure 14 shows how our view of the stars depends upon our position on the Earth. Someone placed at the north pole sees only the northern half of the celestial sphere. The same constellations are always above the horizon, but they are not necessarily visible since, for six months in the year, the Sun is also permanently above the horizon, making the pole a poor

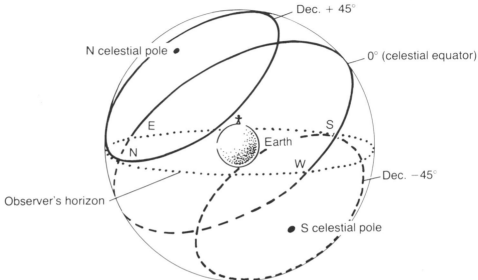

Dec. + 45°

N celestial pole ●

0° (celestial equator)

E

S

Earth

N

W

Dec. −45°

Observer's horizon

● S celestial pole

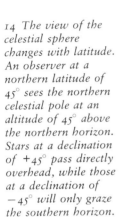

14 The view of the celestial sphere changes with latitude. An observer at a northern latitude of 45° sees the northern celestial pole at an altitude of 45° above the northern horizon. Stars at a declination of +45° pass directly overhead, while those at a declination of −45° will only graze the southern horizon.

site for an observatory! On the other hand, an observer placed at the equator sees every part of the celestial sphere passing above his horizon in the course of one revolution – remembering, of course, that no stars can be seen during the daytime hours. Someone placed at the Earth's south pole would have a monotonous view of the southern half of the celestial sphere, to complement that of his northern counterpart.

Most people live somewhere between the poles and the equator, and therefore their view of the celestial sphere ranges between these extremes. For example, an observer in southern France, at a latitude of 45°N, will at one time or another see all the stars in the northern half of the celestial sphere, and those down to a declination of 45° southern declination (written −45°), or halfway from the celestial equator to the south celestial pole.

How the Sky Moves

The Earth has two important but separate motions. The first is its 24-hour rotation on its axis. This is what makes the celestial objects appear to revolve around the sky, and what the equatorial mounting is designed to counteract. It is often called *diurnal* motion.

The second is the annual motion of the Earth around the Sun. This is shown in Figure 15, and has two very important consequences.

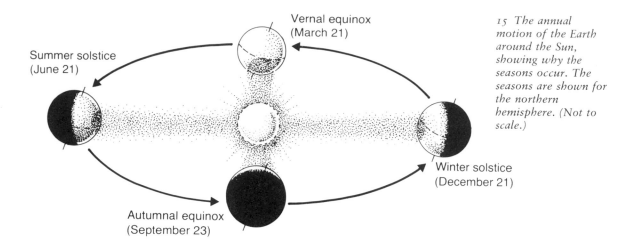

Summer solstice
(June 21)

Vernal equinox
(March 21)

Winter solstice
(December 21)

Autumnal equinox
(September 23)

15 The annual motion of the Earth around the Sun, showing why the seasons occur. The seasons are shown for the northern hemisphere. (Not to scale.)

1 **The seasons** The Earth's axis is not situated at right angles to the plane of its orbit around the Sun, but is tilted at an angle of $23\frac{1}{2}°$ away from the vertical. Since its direction is fixed in space, it follows that the orientation of this axis with respect to the Sun changes during the course of the year. Once a year (round about June 21) the northern point of the axis is inclined at the maximum possible value towards the Sun, and the southern point is turned as far as possible away from the Sun. Such a situation means that the Sun appears relatively high in the sky to dwellers in the Earth's northern regions, and relatively low in the sky to those in the southern hemisphere. This is the time of northern midsummer and southern midwinter respectively.

Six months later (December 21, or thereabouts) the situation is reversed. The southern hemisphere is now tilted towards the Sun, while the northern hemisphere is turned away. Half-way between these two extreme situations, around March 21 and September 23, neither hemisphere is preferentially placed, and the Sun shines equally on each. The two former positions are known as the *solstices*, while the two latter positions are the *equinoxes* – thus, we speak of the summer solstice, the vernal (spring) equinox, and so on.

2 **Star seasons** The stars can be seen only at night. This is not because they are absent from the daytime sky – but the Sun is so bright that it blots all these faint specks from sight. Only the Moon, and sometimes the brilliant planet Venus, can be seen by the normal eye when the Sun is above the horizon.

If the Earth and the Sun were fixed in relation to each other, the stars near the Sun would be forever invisible. However, the Earth revolves around the Sun in its annual orbit, which means that the Sun appears to travel in a continuous path in front of the stars, returning to its original position after a lapse of twelve months. If it lies near a particular star on a given date, it will be on the opposite side of the sky six months later, so that the star now shines in the dark night sky.

For example, one of the best-known constellations in the sky is the group known as Taurus, the Bull. The Sun's annual path makes it appear to pass through this constellation in June of each year, and so Taurus lies in the daylight sky at this time. By December, however, the Sun is diametrically opposite Taurus, and the constellation is well-placed for observation. In fact, being opposite the Sun means that the constellation must be rising at one horizon at the same time as the Sun is setting at the opposite horizon – and setting again as the Sun rises at the beginning of the following day!

This is what is meant by star seasons. The Earth's position in its orbit around the Sun dictates which stars are invisible in the daytime sky and which can be seen above the horizon when night falls. Different times of the year are, therefore, best for viewing different constellations.

The Stars and Constellations

If you look up at the night sky, it will immediately be obvious that the stars differ greatly in brightness. This happens for two reasons: some stars are very much more luminous than others, and send out far more light; and some are many times nearer to our solar system, and therefore appear brighter than other similar stars much further away.

To classify a star's brightness, it is graded in terms of *magnitude*. The brighter a star appears, the smaller is its magnitude value. The brightest star

16 Aldebaran is the brightest star in the constellation of Taurus, the Bull. In December, it appears opposite the Sun in the sky, and can be observed all night. In June, the Sun seems to lie in the same direction as Aldebaran, and the star cannot be seen. (Not to scale.)

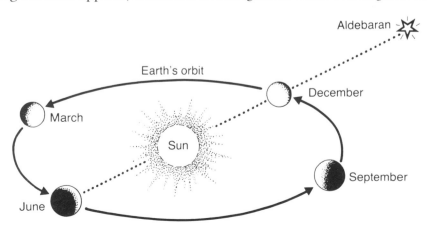

in the sky, Sirius, has a magnitude of −1.6. The faintest stars visible without a telescope, assuming reasonable sky conditions, have a magnitude of about 6.0. One magnitude scale difference means that two stars differ in brightness by about 2.5 times. Two stars differing in brightness by exactly 5 magnitudes have a brightness ratio of 100 between them. Sirius is about a thousand times brighter than a star of magnitude 6.0.

Astronomers sometimes speak in general terms about stars of the '2nd magnitude', or '5th magnitude'. These refer to stars in the range of magnitudes 1.6 to 2.5, or 4.6 to 5.5, and so on.

The *constellations*, of which there are eighty-eight officially-recognized ones, are areas of the celestial sphere named after mythological figures, animals, or even mundane artefacts; near Leo (the Lion), for example, there is a sextant. Some constellation names date back to ancient times, while others are only a couple of centuries old. Few have the slightest resemblance, in their star patterns, to the object they are supposed to represent!

The naked-eye stars in each constellation are given Greek-letter symbols, in approximate order of descending brightness, so that the brightest star is Alpha (α), the second brightest Beta (β), and so on. This designation is followed by the genitive case of the constellation's Latin name. Thus, the brightest star in the Lion (Leo) is Alpha (α) Leonis – also known as Regulus, since many bright and important stars have received special names of their own.

The stars and constellations form the background against which the splendid pageant of astronomy is played. From our moving Earth, we have a participant's view of the universe. Whether your equipment is an astronomical telescope, a pair of binoculars, or just your own two eyes, it is time to launch forth and explore it.

2

The Sun and the Moon

The Sun gives us light and warmth, and without it there could be no recognizable life upon the surface of our planet. It is a hundred times the diameter of the Earth (a million times the volume), and has a surface temperature of about 6000°C, which is about three times as hot as a steel-making furnace. Towards its centre, this temperature must rise to unimaginable heights – many millions of degrees.

The Moon is our satellite. It appears to orbit us once in twenty-nine days, and during this time our view of the hemisphere which is illuminated by the Sun changes in a regular way. The Moon therefore seems to pass through a cycle of phases. It has only a quarter of the Earth's diameter, and astronomically it is one of the less important members of the solar system. Life could continue on the Earth if the Moon were removed, although the ocean tides would almost disappear, and clear nights would always be dark, illuminated only by the stars. It is natural to consider the Sun and the Moon as a duo, since they are the only astronomical bodies to show large discs in the sky. In fact, due to some strange quirk of chance, both *appear* to be about the same size, measuring about half a degree across. But their similarity of size is a great deception, for the Moon is 400 times smaller than the Sun, and also about 400 times nearer to the Earth! You will need to use entirely different observing methods with these utterly disparate bodies, for one is a raging star, the other a rocky satellite.

Observing the Sun

The Sun as a Star

What are stars, and why do they shine? I will return to this question in more detail in Chapter 5, and it will be sufficient to say here that stars are believed to form when huge clouds of 'gust'* begin to collapse inwards and become dense and hot. In the case of the Sun, this probably happened about 4600 million years ago. The central core of the collapse became our shining star, while smaller eddies in the swirling cloud condensed into the cold planets. The planets failed to become fiercely hot and to shine brightly only because there was not enough material inside them – the temperature at the centre of a collapsing cloud depends upon the mass of the cloud, and it has to reach a certain critical limit of some millions of degrees before the nuclear reactions that 'turn a star on' can begin. So, although all the planets became very hot at their centres, they are now essentially cold on their surfaces, and without the light of the Sun they would be dark and invisible.

*The author's suggested term for 'a mixture of gas and dust': the ubiquitous material from which stars and planets have formed.

Since there are countless millions of stars in our own star-city or *Galaxy*, and countless other galaxies in the universe, it might seem as though there must be an amazing variety of star types. In fact, although there certainly are different types of star, they seem to fall into well-defined patterns, and the majority belong to a huge family known as *main-sequence* stars. More or less in the middle of the main-sequence group we find stars like the Sun, and there are millions of them. There are also numerous stars brighter and hotter than the Sun, and very many dimmer and cooler. But we don't find stars that are dimmer and hotter, or brighter and cooler, than the Sun. (Such stars do exist, but they are rare and do not belong to the main sequence.)

So, when we look at the Sun, we are looking at a fairly 'typical' star in terms of mass, brightness and size. It may even be typical, for its class, in having planets revolving around it – we do not know. Neither do we know, though we should dearly like to find out, if planets with intelligent life on them are scattered widely through the Galaxy in which we live. Our interest in the Sun is in its normality rather than in its oddness. It is not surprising that professional astronomy pays so much attention to the Sun. Special satellites have been launched to study it, and day-to-day monitoring of its surface features is carried out by several large observatories around the world.

The Sun's Radiation

The Sun must never be looked at directly, under any circumstances, not even if it is dimmed by thin cloud. The fact that the Sun looks dim does not mean that it is safe, for it sends out dangerous radiations that are invisible to the eye, as well as the blinding ones that hurt at the time.

There are many objects on Earth which illustrate this point. Any glowing body (red-hot coal, electric lamp, or a star) sends out a whole variety of radiation energy. This radiation can be sorted out according to the distance, or *wavelength*, between adjacent pulses. If the distance is very long (say, metres or centimetres), it affects radio circuits. If it is only millimetres or tenths of millimetres, it is felt as heat. If it is around a thousandth of a millimetre, it affects the retina and is known as light radiation. Wavelengths shorter than this come into the group of radiations that include X-rays and other rays that can kill living tissue.

Some stars send out a 'mix' of radiation that contains principally long-wave radiation. These are cool stars, and appear reddish. Very hot, white stars send out a disproportionate amount of dangerous short-wave radiation. The Sun, being an average star, sends out something of everything, and only our atmosphere shields us from the particularly harmful short-wave component of its broadcast energy. Even so, a telescope can concentrate enough into the eye to cause terrible damage. The same is true of the equally invisible heat rays and *infra-red* rays. They are invisible, but present; therefore the Sun's visual dimness is no guide at all to the safety of observing it!

There is one completely safe way of looking at the Sun's surface features: by projection. An image of the disc is cast on to a white screen, and can be viewed with no qualms at all. If you decide that you want to observe the Sun directly through a telescope, it is essential to use a *safe* solar filter, one that has been made for the purpose. It must absorb all dangerous radiations, and not just the visible ones.

Sunglasses are *not safe*: they absorb far too little radiation.

Smoked glass is *not safe*: the carbon deposit is fragile and does not absorb short-wave radiation efficiently.

Coloured glass *Sun-caps*, sold to be placed over the telescope eyepiece, are *not safe*: they do not block out all the dangerous radiation, and may become extremely hot and crack.

Densely-exposed *photographic film* placed in front of the object glass is *not safe*, since it does not block out all the dangerous radiation, no matter how dim the Sun may appear to be.

In short, the only way to make direct solar observations is to use a proper solar filter. The best known is 'Solar-Skreen', available from some astronomical suppliers. It is a limp aluminized plastic sheet, carefully manufactured so as to have no effect upon the quality of the telescopic image. Tests have claimed that this material passes the requirement for safe solar work, and it has been used successfully by solar observers all over the world. Some reputable telescope manufacturers also produce their own filter material, but it would be wise, perhaps, to treat such filters with reserve until you have some independent confirmation that they are perfectly safe.

In fact, no filter is 'perfectly' safe, and this is a point emphasized again and again, quite rightly, by those who advise you not to use a filter under any circumstances, but to rely on image projection. The safest filter in the world can be damaged – even a pinhole can pass dangerous radiation – or fall off during use. It is up to you to take stringent precautions against such eventualities.

How to Observe the Sun

If you are using a *filter*, then fit it over the object glass according to the manufacturer's instructions, and you are ready to begin.

It is enjoyable looking at the Sun with both low and high magnifications. A low power, such as × 30 or × 40, will give a fine view of the whole disc, and any sunspot groups visible can be seen in relation to each other. If you want to look in close-up detail at a particularly interesting group, then turn to a high magnification. With powers of about ×100 or more you often catch glimpses of detail on parts of the surface, or *photosphere*, where there are no sunspot groups at all. This detail is always faint and hard to hold, and is caused by numerous groups of hot cells, the solar *granulation*, which boil up to the surface from the hotter regions within the Sun. If the air is unsteady, they probably will not be seen, and so the ease of visibility of these ghostly patterns and marks is a measure of the quality of the seeing on that particular occasion.

Observing by *projection* involves focusing the image of the Sun on to a white screen. Many new refractors include a solar-projection screen with the telescope, and instructions on how to use it, but you can undertake projection observation using something very simple, made at home.

Solar projection is in many ways similar to using a slide projector to cast a picture on a screen. The picture used in a slide projector is usually a colour transparency, measuring about 24 by 36 mm along the sides, but the projection lens can enlarge this to a much greater size if required. The further the projector is moved away from the screen, the larger the image becomes.

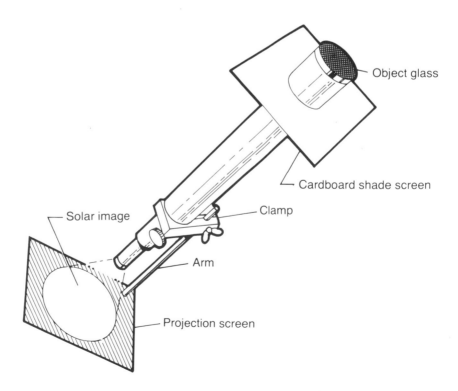

17 How to project an image of the Sun on to a projection screen. The cardboard shade screen casts a shadow on to the projection screen and improves the clarity of the solar image.

Object glass

Cardboard shade screen

Solar image

Clamp

Arm

Projection screen

You can look upon solar projection in the same way. The telescope object glass forms an image of the Sun somewhere in the air, inside the telescope drawtube. If you were to place a small piece of paper near the end of the drawtube, an intensely bright, rather small solar image would be caught – try it and see. The telescope eyepiece then acts like the projector lens, and can be adjusted to cast a sharp enlarged version of this little image on to a screen. The greater the separation of the eyepiece and the screen, the larger the image will be – it will also become dimmer, since the light focused into the primary image inside the drawtube is being spread over a larger area, and diluted in intensity.

Any eyepiece will project an image of the Sun, but if you want to see the whole disc of the Sun at one view, do not use a high-power eyepiece. One giving a magnification of about ×50 will be satisfactory. Experiment, holding a piece of white card about 30 cm away from the eyepiece, and move the drawtube in and out until the image is sharp. The eyepiece will have to be racked out a little from the position you set it in for normal viewing. Move the card to different distances from the eyepiece, and discover how the size of the image changes.

An image diameter of 10 or 15 cm is suitable for most observing. You can draw a circle of the right diameter on the card, or leave it blank, as you wish. There will be more to say about the practical aspects of solar observation in the next section.

A simple mounting for a projection screen can be made from pieces of wire (taken from a couple of old coat-hangers, for example), bent to hold the screen at the right distance from the eyepiece. The other ends can be secured to the telescope tube using rubber bands. It is important to add a counterweight to the upper end of the tube, since the balance will have been destroyed by the addition of the screen. Fortunately, this problem can be solved quite neatly by fitting a cardboard shade screen to the upper end. This is necessary if direct sunlight is to be kept off the image.

Some enthusiasts have made an improved projection system by fitting a projection 'box' rather than a simple screen. A box of thin card, of the appropriate dimensions, can be fitted over the eyepiece, carrying a screen on one of its internal faces. One side is either removed completely, or carries a flap so that the image can be observed. By this means, almost all daylight is excluded from the image, and a much brighter view can be obtained.

The Nature of Sunspots

The Sun is one of the very few astronomical objects which can almost be guaranteed to change its appearance noticeably every twenty-four hours. Although there are occasional periods when no spots may be seen for several days, they are rare. In general, it is exciting to look at the Sun and see what has appeared on its surface.

Sunspots are the surface evidence of tremendous magnetic whirlwinds or vortices down in the solar interior. Energy from the Sun's core, in which particles from hydrogen atoms are being stripped and re-processed to form helium atoms – at an estimated temperature of some 15,000,000°C – gradually works its way to the surface, from whence it radiates into space. These fierce magnetic fields interrupt the flow of energy to the photosphere; where the field cuts through the photosphere, the energy supply is locally reduced. The result is a cool patch – a sunspot – the central part or *umbra* of which is about a thousand degrees cooler than the photosphere. Surrounding the umbra is the lighter *penumbra*. Although relatively cool, the umbra is still hotter than many stars, and it appears dark only by contrast with the much brighter photosphere.

Usually, these lines of magnetic force form an invisible arc above the surface, passing through two points on the photosphere separated by perhaps tens of thousands of kilometres, and producing spot activity around each one. A pair of spots on this pattern is said to be *bipolar*; an isolated spot is described as *unipolar*. However, since the magnetic patterns can be very complicated, you are unlikely to see a pair of perfect individual spots. Most bipolar groups are complex, and consist of numerous detached umbrae and penumbrae. Single spots tend to be neater. Irregular groups, with no clear shape or pattern, are often seen.

The Sun appears to rotate once on its axis in an east-to-west direction in 27.28 days. Therefore, a sunspot group will take about a fortnight to pass from the eastern limb of the Sun to the western limb – if it lasts long enough to be followed for this length of time. The westerly component of a bipolar group, known as the *leader*, usually forms a day or two before the eastern component, or *follower*, and may outlive its companion by days or weeks. The time which must elapse from the first appearance of the leader as a tiny dot or *pore* until the group's maximum development is typically about ten days, and traces may linger on for several solar rotations. Many groups, however, fail to achieve full maturity, and last for only a few days.

The overall appearance of spots is controlled by the mysterious *solar cycle*. Over a period of approximately eleven years, the Sun's magnetic activity rises and falls, and, with this change, the number of sunspots visible

18 The Sun rotates from east to west, taking about four weeks to make one rotation. This diagram shows how some typical spot groups would appear to move across the disc, assuming that their appearance remained unchanged throughout the two-week period.

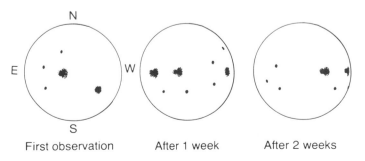

First observation   After 1 week   After 2 weeks

also changes. During minimum activity, which last occurred in 1975–6, 'spotless' days were sometimes observed. At maximum activity, which was last attained in 1980, ten or more groups may be observed simultaneously. There are, therefore, two interesting lines of study: the general solar activity, and the spots themselves.

Observing Sunspots

Sunspots can be beautiful things, and not the least impressive fact about them is their size, for almost any group is going to be larger in diameter than our own planet! The overall diameter of a mature sunspot group can easily reach 100,000 km, and the largest group ever observed, in April 1947, measured some 270,000 km from its eastern to its western extremity, compared with the Earth's diameter of only 12,756 km. These disturbances on the surface of our star are on a truly vast scale.

If you are interested in drawing sunspots, the easiest way to do this accurately is by using the projection method. Project the image of the spot on to a grid of pencil lines, and copy the outlines of the umbra and penumbra on to a sheet of paper placed over a similar grid on a clip-board, so that the grid lines show through. If carefully done, the result will be an accurate copy of the sunspot's shape, and you can fill in the details by direct copying. An equatorial, motor-driven telescope is practically essential for this work, however; otherwise you will spend almost all your time trying to bring the spot back to its place on the grid! You can, of course, try copying the shape freehand, but the result is unlikely to be so accurate. The 'obvious' method of drawing round the outline of the spot on a sheet of paper secured to the screen rarely works well, for two reasons: first, the screen and the

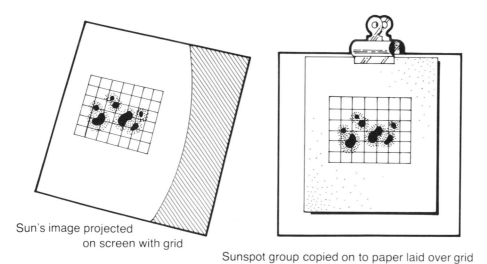

Sun's image projected
on screen with grid

Sunspot group copied on to paper laid over grid

19 How to draw a sunspot group, using a grid of squares to achieve an accurate representation.

telescope tend to shake, and, second, it is difficult to see the pencil line when it is drawn over the image.

If you are interested in drawing individual spots in some detail, you may find that even a 15-cm image is too small for your purposes. If so, either increase the distance from the screen to the eyepiece, or use a more powerful eyepiece, to produce a larger image.

Some sunspot groups change noticeably over the course of just a few hours, and there will almost certainly be obvious changes in any complex group from one day to the next. The main interest in making individual drawings is to follow these changes, and therefore it is particularly important to get the outlines accurate, so that you know the changes are real!

The next step from making individual drawings of sunspots is to draw the whole solar disc, with the various groups represented accurately in outline and position. To do this, you will need to draw a circle on the projection screen which is the size of the solar image, and a similar circle in the observing book or on a sheet of paper. Again, it will be difficult to copy the positions accurately without some sort of grid to guide the eye. (Further information on recording sunspot positions is given in my *The Amateur Astronomer's Handbook* – see list of useful publications.)

One drawback of solar observation is its very convenience – it has to be done during the daytime! It therefore tends to become a 'weekends and holidays' pursuit, since there may be no time to fit it in around work and school hours, particularly in the brief daylight spells of winter. A good drawing, whether of an individual spot on a large scale or of the whole disc on a small one, may easily take a quarter of an hour to achieve, and therefore it may be worth considering other ways of observing the Sun that take up less time and are equally interesting. Of course, there is nothing to stop you simply looking at the Sun and enjoying its turbulent and varied face: but if you don't record what you see, you will have little or nothing over which to reminisce later on.

An alternative to observing sunspots is to make a regular *active area* count. An active area or *AA* is another word for a sunspot, or a group of spots all obviously connected. It may be large or small – a faint pore, barely visible, counts as one AA, as does a large bipolar group. The number of AAs visible on the Sun's face fluctuates on a week-to-week basis, and also on a much longer time-scale, corresponding to the solar cycle. At 'minimum' activity, the disc may be totally devoid of spots for several days at a time, and on these occasions in 1975–6 the AA reading was zero. By the time it had reached maximum activity, in 1980, as many as a dozen or more active areas were waiting to be charted. It is fascinating to start a graph of solar activity and watch it move up and down with the progression of the solar cycle. The next minimum-activity phase is due around the year 1987, and so, if you get started now, you will be well embarked by the beginning of the next cycle!

Other Solar Features

Two features, besides sunspots, will be noticeable on the projected image or in the direct view, provided the image is of the proper intensity. These are the *limb darkening* and the *faculae*.

The limb darkening is visible as a gradual dimming and reddening of the photosphere as the edge or limb of the disc is approached. This is one proof that the Sun does not 'end' at the photosphere, but has a very thin atmosphere cloaking it. Although relatively transparent, this solar atmosphere does absorb a little light. Characteristically, gas absorbs light of short wavelength (the blue end of the spectrum) more severely than it absorbs light of longer wavelength (the red end of the spectrum). Therefore light passing through gas tends to be reddened – this is why the Sun appears red when it is low in our sky, shining through a much greater depth of atmosphere than when it is high above the horizon.

Some features, however, are not dimmed if they lie near the limb; in fact, they may be seen better than when they lie near the centre of the disc. These are the faculae, tangled masses of hydrogen as bright as the photosphere but lying a few hundred kilometres above it. Lack of contrast makes them hard or impossible to make out when near the centre of the disc, but they show up prominently near the limb, since their light, originating higher in the solar atmosphere, is dimmed less than that of the photosphere. Faculae are associated with active areas, marking out a potential sunspot region before any spots appear, and lingering after the group has died and disappeared. It is always worth marking the positions of faculae on your whole-disc drawings, and keeping an eye on the area to see if a sunspot group appears in it.

Eclipses of the Sun

We have already seen that a strange chance of nature has made the Sun and the Moon appear of approximately similar size in the sky. Therefore, if the Moon happens to pass directly in front of the Sun – which it may do at the time of New Moon – the solar disc can be blocked from view. However, the Moon's orbit is noticeably elliptical, so if an eclipse occurs when it is near *apogee*, or at its greatest distance from the Earth, its disc appears too small to cover the Sun completely, and a ring of photosphere remains. An *annular* eclipse, as this is called, is not particularly interesting, since the Sun's beautiful pearly halo, its outer atmosphere or *corona*, bursts into view only if the photosphere is entirely obscured.

In addition to the corona, any plumes of glowing hydrogen, known as *prominences*, rising from the limb, will also be seen as rosy spots or projections around the black edge of the Moon. A total solar eclipse is, therefore, a glorious sight. Although one can never last for more than $7\frac{1}{2}$

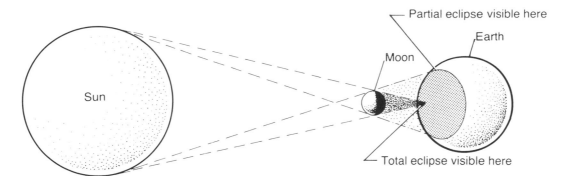

Partial eclipse visible here

Earth

Moon

Sun

Total eclipse visible here

20 *An eclipse of the Sun. The area where a partial eclipse is visible surrounds a very much smaller area where a total eclipse is seen. (Not to scale.)*

minutes, and most are much shorter, enthusiastic amateurs will happily travel a very long way to observe one. This is usually necessary, since a total eclipse can be seen only if you are located within a narrow band, typically a few hundred kilometres wide, that marks the track of the Moon's shadow across the Earth. The following table gives details of the solar eclipses which are due to occur before the end of the century.

Total Solar Eclipses up to the Year 2000

Date	Duration of Totality	Region of Visibility
1987 Mar 29	0^{m}56^s	North Atlantic
1988 Mar 18	3 46	Indian Ocean, East Indies, Pacific
1990 Jul 22	2 33	Finland, USSR, Pacific
1991 Jul 11	6 54	Pacific, Central America, Brazil
1992 Jun 30	5 20	South Atlantic
1994 Nov 3	4 23	Peru, Brazil, South Atlantic
1995 Oct 24	2 5	Near East, India, Pacific
1997 Mar 9	2 50	USSR, Arctic
1998 Feb 26	3 56	Pacific, Central America, Atlantic
1999 Aug 11	2 23	Atlantic, France, Central Europe, India

Observing the Moon

Phases and Features

If you have a telescope, or even a pair of binoculars, the Moon is the most tempting object in the sky. Its regular change of shape is, itself, attractive; even the naked eye can make out some features on the disc, tantalizing enough to demand closer scrutiny with some optical aid. Recent research has brought home to everyone the fact that the Moon is sensationally

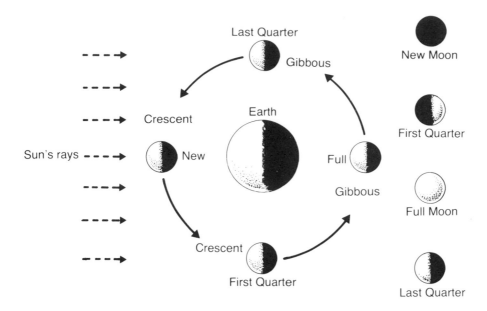

21 *The phases of the Moon, as seen from the northern hemisphere. (Not to scale.)*

different from the Earth, and, in a sense, it has become more intriguing rather than less because of the enormous contrast between the two worlds. As planets go, the Earth is the exception, not the Moon. Mercury and Mars, and the four great satellites of Jupiter (three of which are larger than the Moon) have bare, dead surfaces. We look in vain for any companion, even a remote one, for the Earth. Therefore the Moon is a 'typical' type of small planetary body, just as the Sun is a 'typical' medium-sized star. Its importance lies in its ordinariness.

To us, of course, the Moon is anything but ordinary. It lightens the nights and creates the tides, and is an object of beauty and wonder in our skies. Space travel has not changed that!

The phases of the Moon Why does the Moon pass through phases? As Figure 21 shows, the Sun shines on only one hemisphere, and our view of that hemisphere changes according to the angle between the Sun and the Moon, as seen by the observer. This varies from practically 0° (at New Moon) to about 180° at Full.

At *New Moon* the bodies are arranged so that the Moon is in the same direction as the Sun in the sky (if the line-up is perfect there will be a solar eclipse, but this happens at about one New Moon in six, on average). Normally the New Moon is invisible in the brilliant vicinity of the Sun, its night hemisphere turned towards us. Two or three days pass, and our satellite has moved far enough in its orbit to appear well to the east of the Sun in the sky (to the left of the Sun as seen from regions north of the Earth's equator, and to its right if observed from south of the equator). A narrow sliver of the illuminated hemisphere is now visible: this is the *crescent* phase, when the Moon shines like a graceful arc in the evening sky after the Sun has gone down.

First Quarter arrives a week after the time of New Moon. The Moon is now 90° away from the Sun, and appears as a perfect half (this phase is so called because the Moon has achieved a quarter of its orbit around the Earth). It then enters the *gibbous* phase, and finally shines as a bright, round *Full Moon* after another week has passed. Note that the Full Moon appears opposite the Sun in the sky, so it follows that when the Sun sets, the Moon rises into the darkening sky, and sets again at the following dawn. Therefore, the Full Moon is above the horizon all night, to the annoyance of people who want to observe faint celestial objects.

The Moon then carries on through the second half of its orbit. It will now be high above the horizon only in the later hours of the night, since it does not begin to rise until after the Sun has set, and this interval becomes longer and longer as it approaches another New phase. At the *Last Quarter* phase it is highest in the sky at dawn, and finally the waning crescent, often known as the *Old Moon*, is seen rising not long before the Sun itself. The phase cycle or *lunation*, which lasts about 29½ days, has ended.

When the Moon is a thin crescent, you will notice the rest of the disc (the night hemisphere) glowing against the sky. This effect is known as *Earthshine*, and is caused by sunlight reflected back from the clouds of our own planet, which appears large and brilliant in the lunar sky. Earthshine becomes more and more difficult to see as the Quarters are reached.

The lunar features The Moon has what is known as *captured rotation*. This means that, apart from a small residual swing known as *libration*, it keeps the same hemisphere turned inwards towards the Earth; it therefore spins on its axis in the same time that it takes to orbit the Earth. The cause must lie in the remote past, when it was closer to the Earth and strongly affected by our gravitational field, which 'grasped' one hemisphere and slowed down what was once a rapid rotation. This means that we always see the same presentation of dark seas or *maria* and bright, crisp craters. What changes is the moving belt of contrasting shadow which lies at the borderline between lunar day and night. This is known as the *terminator*, and it sweeps across the surface twice in each lunation. From New to Full Moon we see the morning terminator, where the Sun is rising over the lunar landscape, and after Full the evening terminator betrays the regions where the Sun is setting before the long, chill lunar night.

The terminator is the region of shadow. This is always a glorious sight, and some of its detail can even be seen with binoculars if these are used carefully. The lunar features look so sharp and crisp that it is hard to believe they are natural. Crater walls and central peaks, isolated mounds, long

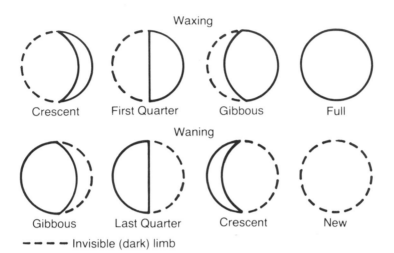

Waxing

Crescent   First Quarter   Gibbous   Full

Waning

Gibbous   Last Quarter   Crescent   New

- - - - Invisible (dark) limb

22 *How the terminator moves across the Moon's surface in the course of a lunation.*

valleys, mountain chains, all have their black shadows attached, and shine in brilliant contrast at the edge of the night hemisphere. Conversely, regions of the Moon well away from the terminator, where the Sun is shining high in the sky and shadows are short or absent, appear as a dazzling patchwork of silver and grey, and it is hard to tell what you are looking at.

You are looking at the remains of ancient convulsions, frozen in time. Astronomers are almost totally agreed now that the early days of the solar system, some 4500 million years ago, were the scene of a destructive sort of creation. Planetary bodies condensed in a whirling cloud of 'gust' surrounding the young Sun. Inevitably, there were collisions, and the larger bodies grew by the accretion of smaller ones on their surfaces. Eventually, the major planets and satellites that we see today had formed, and their surfaces began to cool and harden, but countless smaller solid bodies, up to perhaps a few kilometres in size, remained. If any of these plunged into a 'crusted' planet, it would leave a scar many times its own size.

The lunar craters are scars from collisions with such bodies far back in time, perhaps more than 4000 million years ago. They have been preserved so well because the Moon has no atmosphere and therefore no weather to erode them; also, being a small body, it quickly cooled through and exhibited none of the prolonged geological activity experienced by the Earth to this day. Undoubtedly, our own planet suffered a similar bombardment, but its ever-active crust has thrown up mountains to hide the scars, or they have been lost beneath the sea, or simply eroded away by natural processes, and few crater formations can be detected now.

The dark areas, the maria, are believed to be younger features. Probably they formed when molten rock seeped through the crust from the interior, flooding some of the ancient craters; they are, perhaps, 3000 million years old. Evidence that the Moon was in a 'locked' position even at that epoch includes the fact that the maria are concentrated on the Earth-turned hemisphere, reflecting the gravitational tidal effect of our planet pulling on the Moon's crust. When you look at the maria you will find them relatively devoid of craters, although the ones that are seen tend to be particularly fine examples, since they are late arrivals in the Moon's history, and therefore are far less blurred by later impacts than are the other features.

How to Observe the Moon

Our satellite has something for everybody. Even a pair of binoculars will reveal craters down to about 30 km across, or perhaps even smaller, and therefore many features shown on the overlays accompanying the Moon photographs can be identified. However, if you want to start seeing details inside individual craters, and to have a true sense of being out in space, not far above the Moon's surface, it is necessary to use magnifications of about × 50 or more, and this means using a small astronomical telescope. With a 60-mm refractor, magnifications of from × 50 to about × 150 are useful. If the mounting is very steady, and the air is calm, powers of × 200 can be employed, but do not expect to use such a magnification usefully very often. The best rule is always to use the lowest power that shows sufficient detail, and this applies to any celestial object.

You can go out on any night when the Moon is in the sky and begin to observe it, but a programme of observation will be much more satisfactory. You could, for example, set yourself the task of identifying, and preferably sketching, a list of craters. If you do this, you will need to know how the individual features relate to the phases of the Moon. So the following section describes the objects that are near the terminator, and therefore most distinct because of the shadows that they cast, at different ages of the Moon. The description concentrates on individual craters, valleys, and other compact features, and does not refer to the dark maria unless they are particularly interesting. However, the maria are shown on the outline views, to aid identification of the individual features.

Each view, together with a tracing of the related photograph, shows the phase at which the objects described in the text are near the terminator and well placed for observation. These photographs show the naked-eye or binocular view, not the inverted telescopic view. It will be noted that the orientation shown is not always the same due to the fact it changes as the Moon passes across the sky. Note also that there is no picture of the very slender crescent, which is always hard to observe because it is placed in the sunset sky. Instead, the last map and photograph show the Moon a little after Full, when the features which would be on the terminator in the crescent stage are again on the terminator, and seen to much better advantage because the Moon will now be visible at a good altitude above the horizon.

The 4-day-old Moon

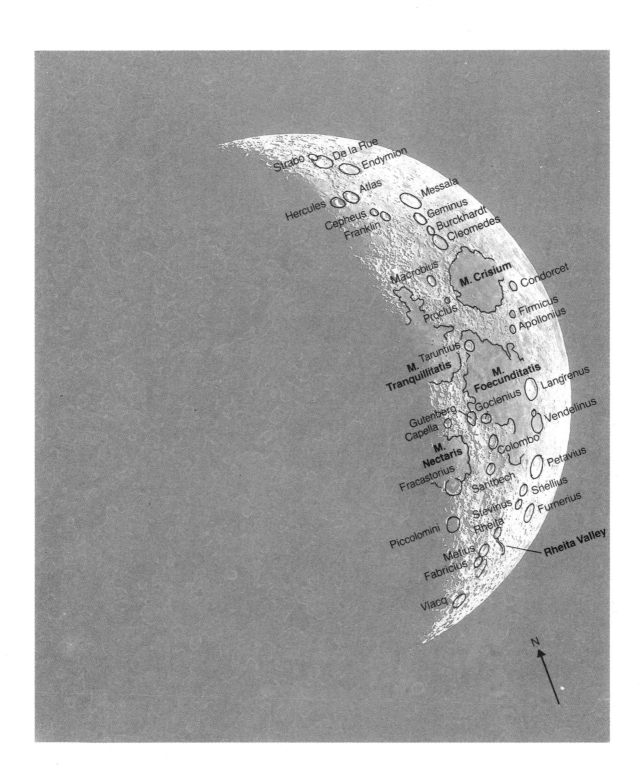

The 4-day-old Moon

The Moon is observable at a younger phase than this, but it may well be very low in the sky, and therefore be seen under poor conditions. Features near the eastern limb are then being observed under conditions of lunar sunrise. On the whole, it is better to observe this eastern region soon after Full, at lunar sunset. The last section (the $15\frac{1}{2}$-day-old Moon) covers objects near the eastern limb.

At 4 days old, the *Mare Crisium* (Sea of Crises) and *Mare Foecunditatis* (Sea of Fertility) are fully lit, and other maria are coming into view.

ATLAS AND HERCULES These craters form a prominent pair, with respective diameters of about 85 km and 70 km. They appear to be of about the same age, being equally well preserved; note the much fainter crater Williams, close to the north of this pair. This must have been formed earlier, and has since been ruined.

FRACASTORIUS This is a most interesting object. Once a splendid crater 90 km across, its northern wall has been reduced by the remelted material of the *Mare Nectaris* (Sea of Nectar), which has flowed into its interior and flooded the original floor, so that it now resembles a mountain-ringed bay.

GOCLENIUS A distinctive 50-km crater in a 'bay' of the Mare Foecunditatis. Close by, to the northwest, is a larger and badly-ruined formation, Gutenberg.

LANGRENUS The northernmost of the 'big three' craters (see also Petavius, Vendelinus), which lie along the eastern limb to the south of the Mare Crisium. It has a diameter of about 130 km, and a very complex central mountain mass. The well-preserved walls also show much detail.

MACROBIUS A splendid 60-km crater, in a prominent position near the Mare Crisium. Its surface is a good reflector of light, making it conspicuous even when shadows are absent.

PETAVIUS This 150-km crater is famous for the wide cleft which runs from the big central mountain group to the southwest wall. This can be seen as a dark line when the Sun is low over the formation, even when a small telescope is used. A day or so after Full is the best time to view this object. It is one of the finest craters on the Moon, and would be magnificent were it nearer the centre of the disc.

PICCOLOMINI A fine crater, 90 km across, lying prominently amongst the surrounding ruined formations of the southern uplands. Its walls are unusually high, and cast long shadows across the floor near the time of sunrise and sunset.

PROCLUS Like Macrobius (see above), a bright crater: less than 30 km across, but very easy to spot at all phases. Notice the bright streaks or *rays* radiating from Proclus. Many craters on the Moon are centres of such ray systems, and they are believed to be the result of silica in the crust being vaporized and literally squirted out by the impacting body. Condensing instantly into minute glassy spheres, they reflect sunlight particularly well. Characteristically, ray craters are young, since subsequent surface-moulding activity would otherwise have obliterated the delicate rays.

TARUNTIUS A fine isolated crater, almost 60 km across. Look carefully at its wall, and you will find a second, lower one within.

VENDELINUS In contrast with Langrenus, this is a badly-ruined formation, measuring about 150 km across. You can infer from this that it was formed much earlier than its great neighbour, and has suffered from the surface upheavals of later impacts with planetary bodies at some early stage of the solar system's history.

The $5\frac{1}{2}$-day-old Moon

The Moon now appears as a thick crescent, and can remain above the horizon until midnight approaches. The Mare Nectaris is fully illuminated, and so is the famous *Mare Tranquillitatis* (Sea of Tranquillity), which is where the first men to set foot on the Moon landed in 1969. The beautiful *Mare Serenitatis* (Sea of Serenity) is almost entirely in morning sunlight.

Compare the position of the Mare Crisium in this view with that in the previous one. It appears noticeably nearer the lunar limb. This is the effect of the slight swinging or libration of the Moon in its orbit round the Earth.

ARISTOTELES A crater measuring 90 km across. It is generally well preserved, although some later and much smaller craters overlap its walls. Unlike many craters of its size, Aristoteles is conspicuous even when the Sun is high above it, since its surface is brighter than the land surrounding it.

BÜRG A 45-km crater, with a deep floor and strongly-ridged inner walls. It lies in the small dark area known as the *Lacus Mortis* (Lake of Death), and is readily detected because of its isolated position.

EUDOXUS A 'companion' of Aristoteles, and also very conspicuous at all phases. It is about 60 km across, but does not have any prominent mountain mass. The impression given by both these craters is that their floors have been remelted and smoothed since their formation.

The 5½-day-old Moon

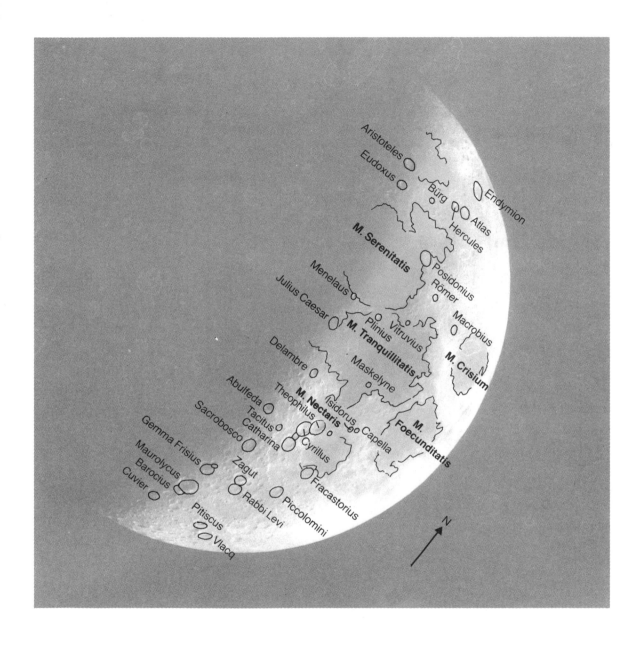

JULIUS CAESAR This must once have been a fine crater, about 60 km across. However, with the passage of time it has been encroached by later craters (the southeastern wall having practically disappeared), and its floor has been darkened by lava flows, so that it now appears as an irregular dark patch when the Sun is high over it.

MASKELYNE A small (30-km diameter) but prominent crater, on the southern border of the Mare Tranquillitatis.

MENELAUS One of the brightest craters on the Moon, this 30-km formation lies on the southern border of the Mare Serenitatis. At Full it appears particularly bright because of the contrast with the surrounding dark mare surface, and a short ray system emanates from it.

PLINIUS This crater marks the confluence of the Mare Tranquillitatis and Mare Serenitatis. It is about 45 km across, very well preserved with high and continuous walls, and evidently is of relatively recent origin.

POSIDONIUS A magnificent crater, one of the finest on the Moon, beautifully situated on the edge of the Mare Serenitatis. Although the walls are not high – presumably having been reduced by later lava flows – they are still in good condition. Many clefts cross the interior. This is an excellent crater on which to practise your drawing skills.

THEOPHILUS, CYRILLUS, AND CATHARINA A prominent row of three craters, all measuring about 80 km across. Theophilus is the best preserved, and has fine sharp walls and a prominent central mountain mass. Cyrillus is badly ruined, and a wide valley connects it with Catharina, the southernmost crater of the three.

VITRUVIUS This crater is only 30 km across, but its dark floor makes it distinctive, since it is sited on a bright 'upland' area.

The 7-day-old Moon (First Quarter)

This is a most impressive phase with any instrument. Many well-placed formations are seen almost in plan view, and with a high magnification there is a feeling of hanging over the surface and seeing the brilliant mountain crests and their black shadows far below you. Part of the narrow northern sea *Mare Frigoris* (Sea of Cold) is in view, and the 'Footballer' can be seen, kicking the Mare Crisium into space!

ABULFEDA This 60-km diameter crater lies on the outskirts of the densely-packed southern uplands, but it is relatively easy to pick out, being better preserved than most of the craters in the area.

ALBATEGNIUS A badly-eroded formation, 120 km across, but in better condition than its huge neighbour Hipparchus (see below). Note the very small central mountain.

ALPINE VALLEY This remarkable straight fault runs through the broad mountain range known as the Alps for about 120 km. At this phase, when filled with shadow, it appears as a black line cutting through the mountains. The Alps themselves are not a particularly lofty range, few peaks exceeding a height of about 2000 m above the general level of the maria.

APENNINE MOUNTAINS This range contains some of the highest peaks to be found on the Moon, and it is beautifully placed for inspection from the Earth, forming as it does an immense curved border to the *Mare Imbrium* (Sea of Showers), which is not yet in sunlight. The mountains extend for some 450 km, and a few peaks rise to about 5000 m above the mare surface. The range can be detected with the unaided eye a little after First Quarter, when the illuminated mountain summits may be seen projecting over the terminator, while their bases are still in shadow.

BESSEL A small but very conspicuous crater, only 20 km in diameter, lying on the Mare Serenitatis. It is located on an isolated bright ray crossing the mare surface. It is worth making an effort to watch the Sun rise over Bessel during the lunar morning, and observe the small inner shadow shrink and disappear as the morning advances.

CASSINI This is a most interesting crater. It measures about 55 km across, and stands in a small bay known as the *Palus Nebularum* (Marsh of Mists), its floor having been remelted and flattened. On this floor are two small but distinct craterlets. The visibility of the smaller of the two is a good test of a modest instrument, since it is only about 3 km across. These inner craterlets make Cassini easy to recognize at all phases.

HIPPARCHUS One of the largest formations on the Moon, but also one of the most obscure. It is an ancient feature, almost obliterated by later impacts, and its outline can be made out only at lunar sunrise and sunset; at Full Moon it is practically impossible to identify. Hipparchus is about 160 km across.

MAUROLYCUS Another example of a badly-reduced crater. This must originally have been a magnificent object over 100 km across, but only the eastern wall is at all distinctive today. The western part of Maurolycus has been completely overlain by later impacts.

SACROBOSCO A very large and ancient ruin, 80 km across, the northern and western walls having been partly destroyed by later impacts. Like many objects in this chaotic area of the Moon, it is hard or impossible to pick out at Full.

The 7-day-old Moon

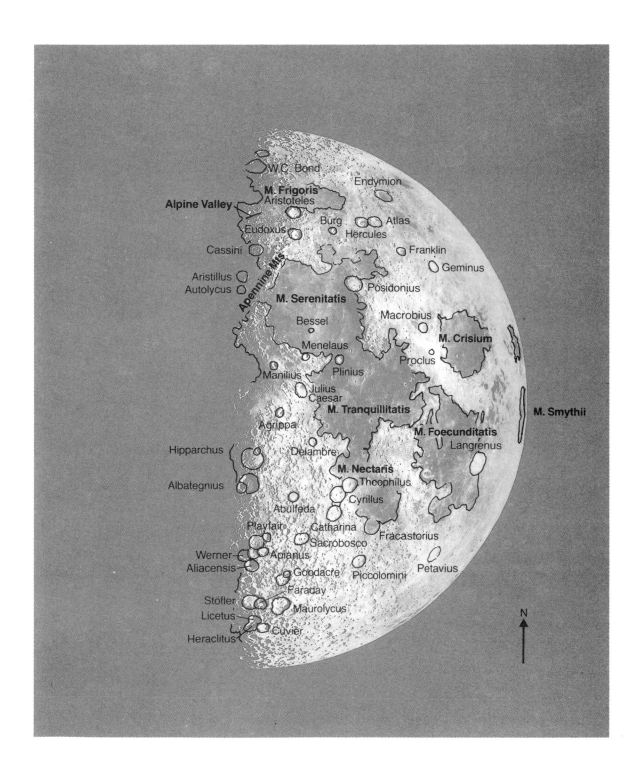

STÖFLER A very large and ancient crater, which must have been an impressive 150-km object in its youth. Now its walls are distorted, particularly by the 45-km crater Faraday, to the east. Stöfler lies in the mountainous southern terrain, where craters overlap and intrude in wild confusion. No doubt the entire Moon once resembled this landscape, but the outflow of hot rock which formed the maria has obliterated huge areas of this ancient surface.

The 9-day-old Moon

Since so many splendid objects come into view at this state of the lunation, some craters visible now will be described in the next section, when they are still well placed. The magnificent Mare Imbrium is almost free of shadow, and the Apennine range is fully displayed. The ray system of Tycho is becoming noticeable.

ALPHONSUS A mighty feature, only slightly smaller than its neighbour Ptolemaeus (see below), and evidently of similar age. However, its original central mountain group remains. The most interesting feature of Alphonsus, which makes it worth observing at all phases of the Moon, is the trio of prominent dark patches inside its walls. These appear to be surface deposits, and they become more conspicuous as the Sun rises higher in the lunar sky – in other words, towards Full. At Full Moon the presence of the crater itself is indicated by these patches, which then appear very dark indeed against the glaring whiteness of the region. Other less obvious patches can be found elsewhere in the crater, notably near the central mountain, and in 1958 a professional astronomer announced that he had observed an escape of luminous gas from this area.

ARCHIMEDES This is one of the group of three craters beginning with 'A' in this eastern part of the Mare Imbrium. It is the largest – about 75 km across – but its floor has been remelted, probably during the formation of the mare itself, and now appears featureless except for some grey streaks. To the south of Archimedes lies an extensive hilly region. No other crater on the Earth-turned hemisphere has so perfect a ring-like appearance as this one.

ARISTILLUS A splendid crater some 55 km across, one of the Archimedes group. There is a central mountain, and the walls are in fine condition, with no subsequent impact blemishes to destroy their regularity. Notice the many hills and craterlets dotted around on the mare surface in this region.

ARZACHEL A neighbour of Alphonsus (see above). It is in rather better condition, and its walls are well preserved around their 90-km diameter.

AUTOLYCUS The smaller brother of Aristillus (see above), 40 km across, but of very similar appearance. These two objects form an impressive sight when on the terminator, for their deep floors hold the shadow for a long time, and they appear like pots of ink.

ERATOSTHENES A fine, moderate-sized crater, 55 km across, with a small central mountain. Like so many of the craters to be found in the maria, it is unaffected by later impacts and its walls are perfectly preserved. It marks the end of the Apennine chain. Two local mountain ranges branch out around Eratosthenes, and it is worth examining the mare surface very carefully when the Sun is low over this part of the Moon, as it has many ridges and bumps which can be seen when shadows are long.

PITATUS An interesting and distinctive 80-km crater. It has been partly overwhelmed by the melting which caused the *Mare Nubium* (Sea of Clouds) but still preserves all of its wall, although only the upper part of the northern arc shows above the mare surface. A trace of the central peak can also be seen.

PLATO A crater familiar to all lunar observers, curiously sited at the western end of the Alps, on the northern shores of the Mare Imbrium. It lies in a lunar latitude of 52° (corresponding almost exactly to the latitude of London on the Earth's surface), and we therefore see its outline as appreciably elliptical. In size (90 km across) and general appearance it strongly resembles Archimedes (see above), for its floor has been remelted and appears dark, like that of the nearby mare. This floor contains a few light patches, and some very small craterlets that are almost certainly beyond the range of a 60-mm instrument.

PTOLEMAEUS This lies on the edge of the southern uplands, very near the centre of the disc, the realm of huge and ancient formations. This is one of the largest, 140 km across, and is badly pitted by later impacts, although its inner walls are still in relatively good repair. It gives the impression of a circular depressed plain. We have an 'overhead' view of this great formation, which contains a 7-km craterlet, Lyot, northeast of the centre. Note the much more recent and well-preserved 45-km crater Herschel, immediately to the north.

TIMOCHARIS A young 40-km crater in a lonely site on the Mare Imbrium, and always easy to identify.

TYCHO A well-preserved 85-km diameter crater, with fine walls and a prominent central mountain mass. In most aspects it resembles Copernicus (see below), although its position in the southern uplands does it less justice, since it does not stand out so prominently from its surroundings. Undoubtedly it is a young formation: not only is it virtually unmarked by later impacts, but it has thrown out a huge system of rays – by far the most extensive on the lunar surface – which have not been obliterated by later surface moulding. One of Tycho's rays stretches into the *Oceanus Procellarum* (Ocean of Storms), a distance of a thousand kilometres or more, and at Full it seems to be the centre of a huge fan of bright beams.

The 9-day-old Moon

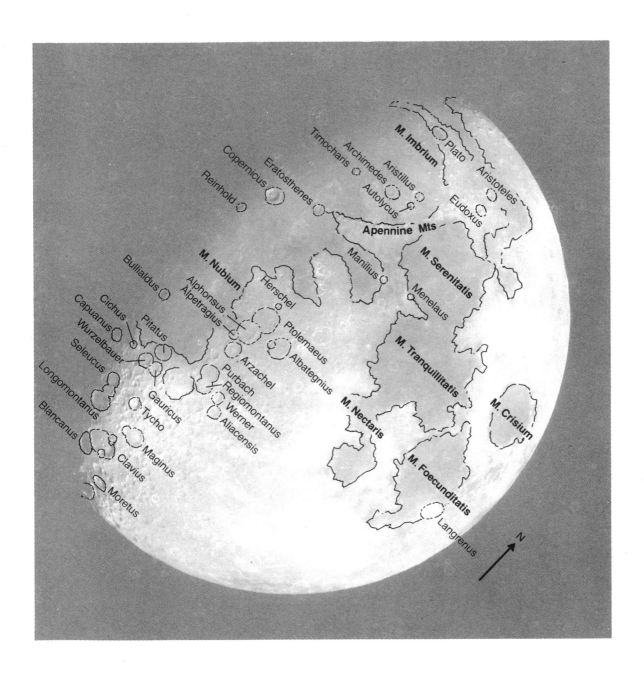

The 10-day-old Moon

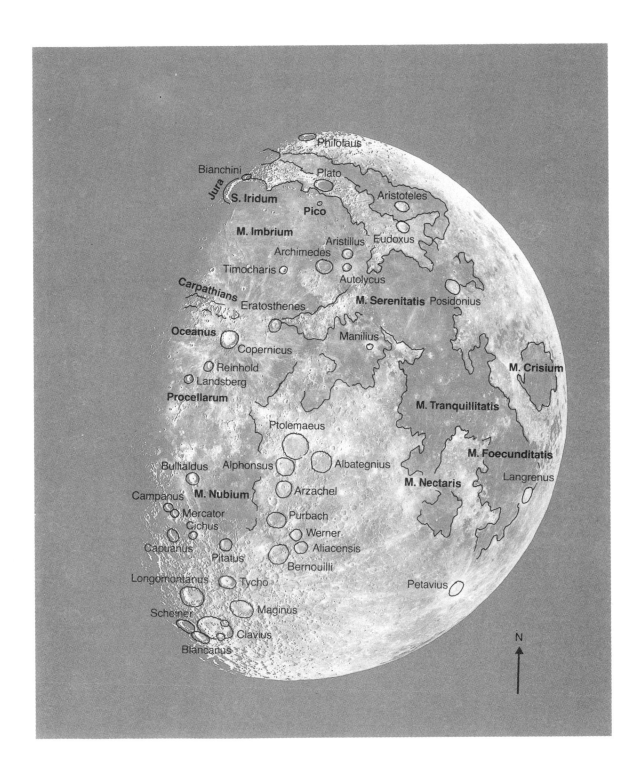

The 10-day-old Moon

The terminator is now well advanced towards the western limb, and a huge area of the Oceanus Procellarum has come into view. It is dawn over the *Sinus Iridum* (Bay of Rainbows). It is worth spending some time examining the terminator region where it sweeps across the Oceanus Procellarum: at first sight it may appear flat and uninteresting, but closer examination should reveal numerous ridges and swellings, tiny peaks, and craterlets.

BIANCHINI A small but distinct 40-km crater on the crest of the mountain ridge forming the Sinus Iridum.

BULLIALDUS A smaller model of Copernicus (see below), 60 km across, with the same detailed, well-preserved walls and a fine central mountain.

CAMPANUS AND MERCATOR A pair of 45-km craters, their walls almost touching. Campanus is the better preserved of the two, and has a small central mountain. This region contains a number of long clefts, noticeable under low-light conditions.

CLAVIUS A colossal formation measuring some 220 km across. Despite later impacts, which have imprinted several fair-sized craters on its face, it is still clearly recognizable, and its inner walls are high and detailed. Note another majestic formation close by to the northeast – Maginus.

COPERNICUS A superb crater, 85 km across, situated at the centre of a prominent ray system, whose general position can be made out with the unaided eye. Its finely-terraced walls and complex central mountain mass make it a magnificent sight when near the terminator, and it is always a conspicuous object, even at Full Moon. Note the mountains to the north, the *Carpathians*, which form a natural boundary between the Mare Imbrium and the Oceanus Procellarum.

LANDSBERG A small crater, 50 km across, prominent because of its isolated position on the Oceanus Procellarum. Its inner walls are terraced, and there is a small central mountain.

LONGOMONTANUS One of the very large craters of the far southern highlands. It is about 140 km across, and is in a reasonably good state of preservation, although a moderate crater has impacted on its southern wall, and the northern part is badly broken.

PICO A lonely mountain on the Mare Imbrium, 100 km south of Plato (see p. 63). This is always conspicuous as a bright spot, but near the terminator it casts a long shadow, since it is some 12,000 m high! Part of the attraction of this region is the fact that we see it at a noticeable 'slant', as though flying past it in a spacecraft, and this adds a touch of extra realism to the telescopic view.

REINHOLD A 50-km crater lying betweeen Landsberg and Copernicus (see above), and strongly resembling the former, even down to the small central mountain.

SCHEINER This splendid 110-km crater lies to the north of Blancanus. It is still well preserved, although the original central mountain has disappeared, and small craters have disturbed the walls, which still retain much of their original terracing.

SINUS IRIDUM The Bay of Rainbows is a remarkable feature, having the appearance of a huge semicircular cliff, known as the Jura Mountains, overlooking the dark mare surface below. Some of these peaks rise to at least 4 km in height, and when caught on the terminator they may be seen with binoculars, or even with the naked eye, as a projecting loop of brightness curving into the night side of the Moon. At such a time, the Sinus Iridum is one of the great sights of the lunar disc.

The 12-day-old Moon

The most conspicuous 'new arrival' on the lunar disc is the brilliant crater Aristarchus, but there are also several interesting formations along the southern half of the terminator.

ARISTARCHUS A 45-km crater, the most reflective on the Moon's surface, and the centre of a small but brilliant system of bright rays. Evidently there is some curious feature about the surface deposits of Aristarchus which accounts for this unusually high reflectivity. It is common for this crater to be visible as a hazy spot under Earthshine conditions. Some dark bands extend from the floor of Aristarchus up its inner walls, and seem to become more prominent as the Sun rises over the formation, but they are hard to see with an aperture of less than about 75 mm.

GASSENDI This superb formation, 85 km across, could not be confused with any other on the lunar surface. We see a mountain ring, rather low, with a central peak poking up out of the flooded floor, and a prominent crater, Clarkson, overlapping its northern wall. It stands guarding the entrance to the *Mare Humorum* (Sea of Vapours), and is a wonderful sight when the terminator passes across it. If you have a telescope with an aperture of 100 or 150 mm, you should be able to make out some of the famous clefts which seam the floor of Gassendi.

HAINZEL This irregular object is a merging of two moderate craters, and stands out because of its 'waisted' shape.

HERODOTUS AND SCHRÖTER'S VALLEY Herodotus is a 35-km crater lying closely west of Aristarchus (see above). Schröter's Valley, the most famous cleft on the Moon, originates at its northern wall, and then turns westwards

The 12-day-old Moon

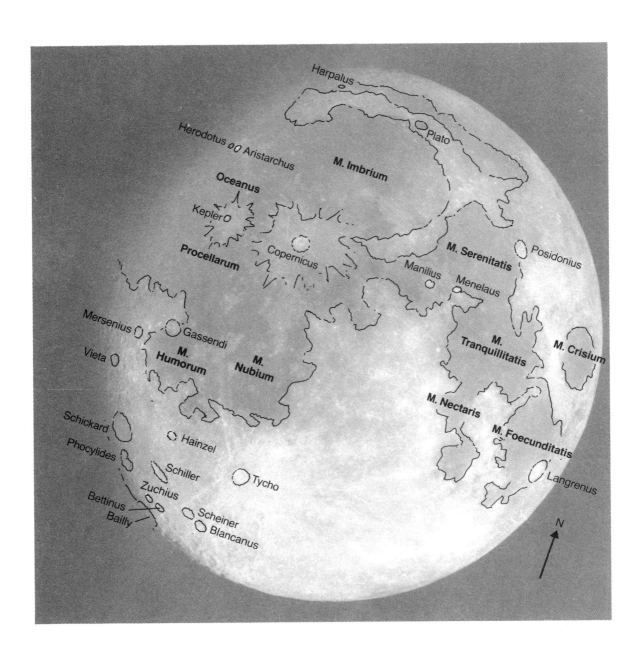

The Full Moon

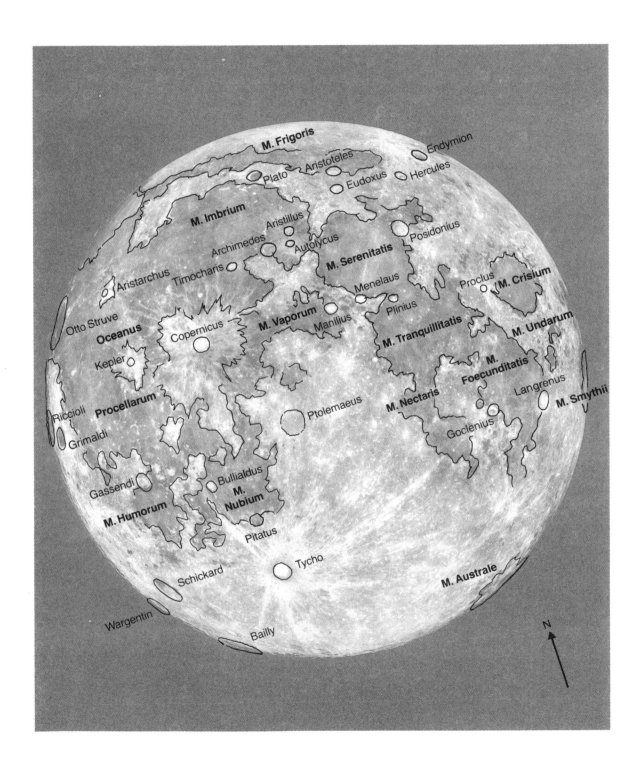

and northwards before petering out. It can be seen as a hairline marking with a very small instrument.

KEPLER A bright ray-crater. It measures only 35 km across, but the high reflectivity of its surface makes it a conspicuous object at all phases, and near Full Moon its rays merge with those of Copernicus.

MERSENIUS An 80-km crater on the western edge of the Mare Humorum, noticeable because of its rather bare interior. A long ridge lies between it and the mare surface.

PHOCYLIDES A 100-km crater with walls rising in places some 3 km above the floor. A smaller but still considerable crater, Nasmyth, abuts it to the north, and gives the feature an irregular appearance.

SCHICKARD An immense, ancient object, over 200 km across. Its floor has been remelted, and appears relatively smooth, with no trace of a central hill. There are some craterlets and other details, however, so that it is worth studying closely. The northern wall has been badly broached by later impacts.

SCHILLER A very curious formation, which has the appearance of an elongated crater some 190 km long but only half as wide – allowing for its inclined appearance, near the limb. It is, presumably, a double crater whose adjoining walls have disappeared.

WARGENTIN Readily located a little southwest of Schickard, and best seen about a day before Full Moon, this is a crater whose floor is *higher* than the surrounding country, and which, in fact, extends in places to the summit of its walls. Therefore you will find little internal shadow as the Sun rises over it. Evidently the impact crater became flooded with molten rock at some time after its formation. (See Full Moon map for location.)

The 14-day-old Moon (Full Moon)

In general, Full Moon is a poor time to observe, since the disc appears shadowless, and the lunar features do not appear in relief. However, a few formations very near the limb can only be well seen around this time, and the following four objects near the western limb come into this category.

BAILLY The largest recognizable crater on the Earth-turned surface, but so near the limb that it is not easy to distinguish. Its diameter is some 290 km, and it is partly obliterated by much smaller overlapping objects.

GRIMALDI A very dark crater on the western border of the Oceanus Procellarum. It measures almost 200 km across, and is always distinctive, even though its wall has been severely reduced by later impacts.

OTTO STRUVE A colossal formation lying very near the northwestern limb, libration making it appear to swing very noticeably towards and away from the edge in the course of the lunar month. From north to south it measures some 300 km, but evidently it is the confluence of two craters rather than the result of a single impact. This is an interesting challenge, since our view is so inclined.

RICCIOLI A dark crater near Grimaldi (see above). It is another ancient object, 160 km across. Its floor is unusually sombre, making it conspicuous at all phases.

The 15½-day-old Moon

As explained on page 54, objects near the Moon's eastern limb are seen under their morning illumination when the Moon is a thin crescent visible in the evening sky after sunset, but the combination of low altitude and twilight may mean a poor telescopic view. A day or two after Full, these craters may be observed much more conveniently, with the Moon above the horizon for most of the night.

BURCKHARDT A small crater, 55 km in diameter, of rather irregular appearance and with high walls. It occupies a conspicuous position, to the north of Cleomedes (see below).

CLEOMEDES A fine, very distinctive crater, 125 km across, easily identified because it lies closely north of the Mare Crisium. It must be an old formation, since its walls are broken in places, and the floor contains no central peak, although the walls rise high above it.

DE LA RUE The remains of an ancient crater, some 150 km across. It has been so overlapped by subsequent impacts that the walls have descended almost to floor-level in places, but it is still easily made out when the terminator is near.

ENDYMION A fine object, with well-preserved walls, 130 km across. Its dark floor makes it distinctive even when the Sun is shining directly down on it, as at First Quarter. It has overlapped another crater of similar size, giving it a double appearance.

FURNERIUS Another fine large crater, 120 km across, but not standing out particularly well, since the walls do not rise very high above the surrounding ground.

GEMINUS A well-preserved crater about 80 km across, with a small central mountain.

MARE CRISIUM This is a particularly beautiful and impressive lava plain, since it is surrounded by mountains, and these cast conspicuous shadows

The 15½-day-old Moon

when the terminator is near. It can be seen easily as a dark spot with the naked eye. The north-south diameter is about 450 km. Even a small instrument will reveal three small craters on its surface, near the western margin: the smallest is about 10 km across. Note the winding ridges crossing this mare, and spend some time looking for tiny craterlets and hills, of which there are many on its surface.

RHEITA VALLEY A rift, huge by earthly standards, measuring about 180 km in length and some 25 km in width in places. In reality it is more like a shallow scoop than a valley. It is not easy to detect, since it lies near the limb in the mountainous southern uplands, and is partly concealed by surrounding craters.

SNELLIUS An 80-km crater, readily picked out on the southern border of the Mare Foecunditatis. The walls are well preserved, and it has a small central mountain.

STEVINUS The southern companion of Snellius, of very similar size and appearance, although its floor is rather darker, and the central mountain is more prominent. It lies close to Furnerius (see above).

Eclipses of the Moon

When the Moon is Full, it is more or less opposite the Sun in the sky. If the line-up is perfect, it passes through the long shadow that the Earth casts into space; direct sunlight is cut off, and the disc turns a deep copper tint.

Only about one Full Moon in eight experiences an eclipse, since the Moon's orbital plane is tilted with respect to our own, and the Moon usually passes north or south of the Earth's shadow. When a total eclipse happens, however, the phenomenon lasts for several hours (the total phase can last for an hour and a half), unlike the lightning rapidity of a total solar eclipse. Also, since a lunar eclipse appears identical wherever it is viewed from the Earth's surface, the chances of observing a total lunar, as opposed to a total solar, eclipse from a given site are several hundred to one.

If a lunar eclipse is due, look up the times of the *contacts* well in advance, so that you are prepared for what will happen. The contacts are the times when the edge of the Moon touches the edge of a shadow. There are two concentric shadows cast by the Earth – the outer one being the very extensive *penumbra*, the faint annulus in which the Sun is only partly obscured. In the centre of the penumbra is the *umbra*, where the Sun is totally obscured. In theory, the umbra should be pitch black, but the Earth's atmosphere refracts light (principally red light) into the shadow, and this gives the eclipsed Moon its characteristic hue. It is worth noting that, if the eclipse is only partial, the red tint may not be noticeable because of the dazzling uneclipsed portion of the Moon's disc, and the shadowed sector looks completely black.

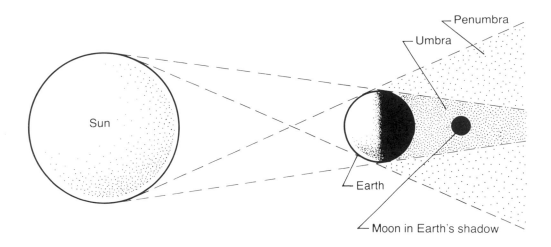

Penumbra

Umbra

Sun

Earth

Moon in Earth's shadow

23 An eclipse of the Moon. (Not to scale.)

The pattern and timing of a lunar eclipse is as follows:

First contact with the penumbra Although this is not detectable, since the initial dimming is so slight, the penumbral dimming should be noticeable after half an hour or so as a hazy shading at the eastern side of the disc (the left-hand side as observed from the northern hemisphere).

First contact with the umbra This will probably occur about one hour after the penumbral contact. The shadow suddenly becomes noticeably darker, and in a few minutes a 'bite' can be seen with the naked eye.

Totality, or second contact with the umbra, will occur about another hour after first contact. By this time, the Moon will have assumed its reddish hue, unless the eclipse is unusually dark. On some occasions, the Moon has almost disappeared from view – this happened in 1963, for example. The cause of these different intensities appears to be changes in the transparency of the Earth's atmosphere, due to emission of volcanic dust or other natural pollution.

After totality, the eclipse stages pass in the reverse order. The sky lightens, and the faint stars that peeped out during totality disappear as the blaze of the Full Moon returns

To watch a lunar eclipse, it is best to use a low-power eyepiece, so that the whole Moon can be seen at one view. The edge of the shadow is very blurred, and little is to be gained by using a powerful eyepiece. Try making timings of the moment when the shadow engulfs selected craters, and make notes of how the very bright craters (such as Aristarchus) and the dark ones (such as Grimaldi) appear when in the shadow. As we have seen, lunar eclipses are not particularly common – so make the most of any that do occur!

3

The Planets

The solar system contains nine major planets, including the Earth. They are called *major* to distinguish them from the countless small bodies, known as *asteroids* or *minor planets,* which circle the Sun in their thousands. A few of these are bright enough to be seen with binoculars, but the majority are very dim indeed, and do not appear in the amateur astronomer's observing list.

Of the nine major planets, one (the Earth) can always be seen by looking down, while five others (Mercury, Venus, Mars, Jupiter and Saturn) are readily detectable with the unaided eye if they happen to be in the night or twilight sky. Uranus, the planet beyond Saturn, is technically visible without optical aid, but is extremely faint, while Neptune can be seen with binoculars, and Pluto requires a fairly powerful telescope for its detection.

The Movements of the Planets

Seen in the sky, planets look just like stars. So how do you tell them apart, without a telescope? By doing what the ancient observers did – noticing that they seem to move relative to the stars. They do not belong to the fixed constellation patterns, but follow strange courses of their own, which are the combined result of their orbital motion and our annual revolution around the Sun. One year, Jupiter may shine in the constellation Leo; the next, it may be in Cancer, and so on. If you watch closely, its motion around the sky will be noticeable on a night-to-night, or certainly on a week-to-week, basis. (Beware of loose terminology. 'Around the sky' means motion relative to the background stars, not to trees or houses. This second motion is the daily rising and setting effect, which occurs because the Earth spins, and is known as diurnal motion.)

Therefore, a star map is of no help in locating the planets, except in one special way. The orbits of all the major planets lie fairly accurately in a common plane (only those of Mercury and Pluto are inclined at an angle of more than $3°$ to that of the Earth), which means that they do not wander all over the sky, but remain in a band that circles it. This band is called the *Zodiac*, and it is centred on the ecliptic, which is, as we have seen, the plane of the Earth's orbit, and which is also the apparent path of the Sun around the celestial sphere during the course of the year. The Zodiac passes through the twelve Zodiacal constellations, and this is where the major planets (and many of the minor planets, too) are to be found.

When thinking about the way in which the planets appear to move around the sky, it is vital to remember that we are observing them from our own planet, which shares in the general motion around the Sun (in an anti-clockwise direction if viewed from an imaginary point far to the north of the orbit plane). If you simply want to observe a planet, it isn't necessary to understand anything about the way it moves; you simply need to know where it

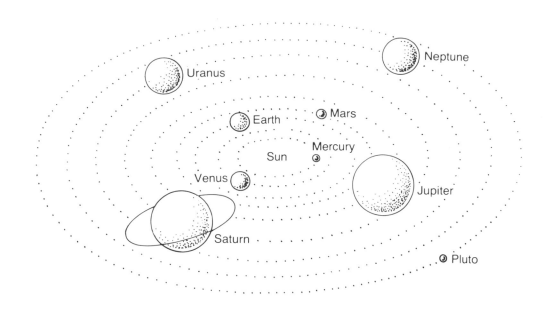

Planet	Mean Distance from Sun (millions of km)	Orbital period	Rotation period	Diameter (kilometres)	Mean surface temperature C	Main atmospheric constituents	Number of satellites
MERCURY	57.9	88 days	59 days	4,878	$350°-170°$	None	0
VENUS	108.2	225 days	243 days	12,104	$480°$	Carbon dioxide	0
EARTH	149.6	365 days	23h 56m	12,756	$22°$	Nitrogen, Oxygen	1
MARS	227.9	687 days	24h 37.5m	6,794	$-50°$	Carbon dioxide	2
JUPITER	778.3	11.9 years	9h 55.5m	142,700	$-150°$	Hydrogen, Helium	14 +
SATURN	1,427	29.5 years	10h 14m	120,800	$-180°$	Hydrogen, Helium	15 +
URANUS	2,870	84 years	24h?	52,000?	$-210°$	Hydrogen, Helium, Methane	5
NEPTUNE	4,497	165 years	22h?	48,000	$-220°$	Hydrogen, Helium, Methane	2
PLUTO	5,900	248 years	6.4 days	3,000	$-230°?$	Methane	0

24 *The normal order of the planets, in increasing distance from the Sun. Note that Pluto, because of the elongated nature of its orbit, will be slightly closer to the Sun than Neptune until 1999. (Not to scale.)*

is in the sky. However, there are some basic facts about the way the planets appear to weave around the sky which *are* important.

The Inferior and Superior Planets

Two planets, Mercury and Venus, are closer to the Sun than is the Earth. The rest revolve around the Sun in much larger orbits. Figure 24 gives a view (not to scale) of the orbits of the major planets, and it will be clear from this that there is one very important difference between our view of Mercury and Venus on the one hand, and the rest of the planets on the other. This is because the other six planets can at times appear more or less opposite the Sun in the sky (a position known as *opposition*), while the two inner ones cannot, and are tied to the location of the Sun. We call these two the *inferior planets*, the others the *superior planets*.

If you imagine the Earth's orbit as one of the grooves in a record, and the other planetary orbits as other grooves, this may help to clarify what happens. Imagine that the Earth is a fixed speck of dust, and that you are somehow attached to it. Set the planets revolving on their way. Looking towards the central hole (the Sun), you will see Mercury and Venus apparently swinging out from the Sun and back again, first on the right and then on the left. Occasionally, too, a superior planet will pass your view beyond the central hole, gradually moving right round the record until it is directly behind you as you face the Sun, and then moving back to the far side of the Sun again.

The Inferior Planets

Let us take a closer look at these two fitful worlds. Hold up your outstretched hand at arm's length. The span of the hand from the tip of thumb to little finger is about $20°$. When Mercury is at its greatest possible angular distance from the Sun, a position known as *greatest elongation*, it is only $27°$ away – not much more than a hand's stretch. Venus, at elongation, appears almost twice as far from the Sun – about $47°$ away. But at times other than elongation they are much closer to the Sun in the sky, being lost altogether from view as they pass its dazzling disc at *inferior conjunction* and *superior conjunction*, as shown in Figure 25.

So when you think of observing an inferior planet, this generally needs to be done at twilight. Mercury always rises or sets within a couple of hours of the Sun, and Venus within about four hours.

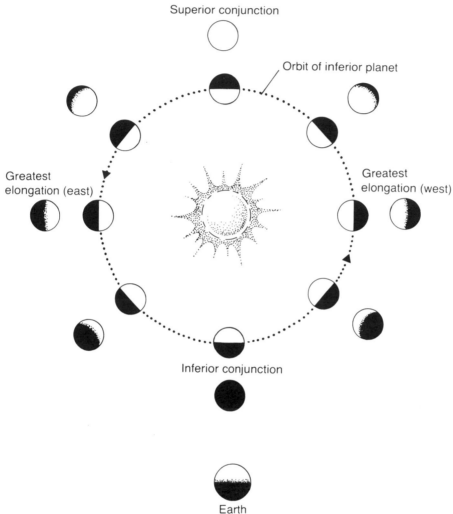

Superior conjunction

Orbit of inferior planet

Greatest
elongation (east)

Greatest
elongation (west)

Inferior conjunction

Earth

25 *How Mercury and Venus appear to move, as seen from above the Earth's northern hemisphere. The outer circle shows the view of the planets from the Earth at each stage of their orbit. (Not to scale.)*

Mercury

Mean solar distance: 57.9 million km (orbit very eccentric; ranges from 46 million to 70 million km).

Length of year: 88 days.

Average interval between elongations: 116 days.

Diameter: 4878 km.

There is a popular belief that Mercury is faint and difficult to see, and you will almost certainly, in your astronomical reading, come across the statement that the great astronomer Copernicus never managed to see it at all. However, Mercury isn't as elusive as all that; at times it can shine as brightly as the very brightest stars in the sky, and be perfectly well visible with the naked eye – the secret of observing Mercury successfully is to know when these times are going to occur. They are the times of elongation, when the planet appears at its greatest angular distance from the Sun.

When the planet is to the *west* of the Sun (to the right, as observed from the Earth's northern hemisphere), it rises in the *morning* sky, before daybreak. When *east* of the Sun, it is visible in the *evening* sky, after sunset. You will find information about morning and evening elongations in the monthly newspaper columns, or you could get advance warning from one of the almanacs described at the end of the book. However, not all elongations, even though their angular distance from the Sun may be the same, are equally favourable. At some seasons, Mercury appears much lower in the sky than it does at others. The most favourable time for observing a morning elongation is in the late summer/early autumn months, while evening elongations are best seen in late winter or early spring.

Hunt Mercury with binoculars. Find a site with a clear view of the appropriate horizon, and await your opportunity. Typically the best time is about $\frac{3}{4}$ to 1 hour away from sunrise or sunset, but the rapid twilight of equatorial regions means that you may be able to observe closer in time than this. It is a question of combating the two opposing evils of a twilight so bright that the planet cannot be seen, and a planet so low in the sky that it cannot be made out in the horizon haze. In the morning, Mercury is getting higher and more favourably placed, but the sky is brightening; in the evening the reverse is happening.

But if you are patient and choose your moment, you will come across a 'star' twinkling at the edge of the twilight glow – a beautiful sight indeed. For a moment you may wonder if it really is Mercury, or just a bright star, but if the major constellations are now familiar you will know the positions of all the stars that are as bright as Mercury at elongation – magnitude roughly from −0.5 to +0.5, depending upon circumstances (Mercury appears fainter if elongation occurs when the planet is near *aphelion*, or farthest from the Sun, than at *perihelion*, when it is closest, since the intensity of sunlight upon its surface is much less on the former occasion).

If you have a telescope, then you will hurriedly bring Mercury into view. Be prepared for a great disappointment: with a 60-mm refractor, × 100, you will be fortunate to see anything but the tiniest pink-white blur. At first sight, you might even think it *is* a star! The reason is that Mercury always appears very small, and is usually observed when low in the sky, so that bad seeing conditions make the image swirl and distort. Its elongation diameter is about $7''$ arc* – corresponding to the apparent diameter of a lunar crater

*$7''$ arc is short for 7 *seconds of arc*. A second of arc is equal to a 360th of a degree, or a sixtieth of a minute of arc ($1'$ arc). The symbols $'$ and $''$ for minutes and seconds of arc are no longer 'approved', but I find their substitutes *arcmin* and *arcsec* unacceptable, and prefer to run the risk of these symbols being mistaken for feet and inches!

about 10 km across, which is smaller than almost anything I have described so far. Mercury's bare, crater-ridden surface almost defies effective observation from Earth, and only in the last twenty years was it discovered (by radio observations) that its 'day' is as long as fifty-nine of ours.

Once you have found Mercury, you will want to follow it from night to night, to see how long you can keep it in view. Unless you live near the equator, you will do well to follow it for longer than a fortnight or so – it rapidly vanishes back into the Sun's rays, to reappear at the opposite elongation some six weeks later. However disappointing as a telescopic spectacle, you will never forget your first sighting of the innermost planet.

Venus

Mean solar distance: 108.2 million km (orbit almost circular).

Length of year: 225 days.

Average interval between elongations: 584 days.

Diameter: 12,104 km.

Venus is much larger than Mercury – it is only very slightly smaller than the Earth – and it is covered with clouds, which are very good reflectors of light. It appears second only to the Moon as a night-time object, and there is never any difficulty in identifying this planet when it shines in the morning or evening sky. With a maximum magnitude of − 4.4, it has been seen to cast a shadow on those occasions when it blazes out of a dark sky, which it can do near greatest elongation.

The size of Venus in the sky varies tremendously from inferior to superior conjunction, and it passes through a complete cycle of phases, which can be followed with a very small telescope. Figure 25 explains what happens. When it is near inferior conjunction, the thin crescent phase can be seen beautifully with binoculars, particularly if you catch the planet in a bright blue sky, before it begins to blaze and mask detail. A 60-mm refractor will show most of the phase cycle, although near superior conjunction the planet appears not much larger than Mercury. Make drawings of the phases, and keep an eye out for any faint shadings that may mark details in the clouds. These will probably elude a small telescope, but you will doubtless notice that the terminator area appears slightly dull compared with the limb.

It is of little use trying to observe Venus when it shines brilliantly in the sky. It will look like a searchlight viewed through swirling vapours. Instead, look for it when the Sun is on the horizon. It is not difficult to sweep up in binoculars, provided of course you have some idea of its position: a white pinprick of light will sail into the field of view, and this is always a satisfying, even thrilling, moment. When Venus is at its brightest (in the thick crescent stage) it has been seen by many people with the naked eye in full daylight.

Finding Venus in daylight In fact, one of the most interesting challenges set by the planet is to pick it up in full daylight. The most convenient time to try to do this is when the planet is on the meridian (i.e., due south to an observer in the northern hemisphere, and vice versa in the southern). Using the *Astronomical Almanac* or *Reed's Nautical Almanac*, or some other publication readily available in a reference library, find out the right ascensions of Venus and of the Sun. If that of Venus is the larger of the two, then it lies to the east of the Sun in the sky; if smaller, it lies to the west.

The following method is appropriate when the planet is *east* of the Sun – in other words, an 'evening star', setting after dusk. Find the difference in right ascension, in hours and minutes. At around the time of local noon (the exact moment is not critical, but within an hour or so is best), take note of where the Sun is in the sky. Fix its azimuth by locating some marker directly below it – this could be a distant tree, or a house, or even a lamp-post – and estimate its altitude by the simple 'outstretched hand = 20°' rule. Take care not to look at the Sun itself when doing this.

If you now wait for a time interval equal to the difference in their right ascensions in hours and minutes, Venus will have the same azimuth as you have noted for the Sun. Take a pair of binoculars and sweep the sky upwards and downwards, shifting a degree or two in azimuth after every other sweep and repeating the process. If the sky is a good deep blue, the planet should be found quite easily, as long as it is more than about 15° from the Sun – the bright atmospheric halo that the Sun generates around itself can make Venus rather tricky to find near the time of conjunction. One important, indeed vital, precaution: ensure that the binoculars are accurately in focus for a distant object. If the planet is even slightly defocused, it may well slip through the field as an unperceived ghost.

If the planet is a 'morning star', lying to the west of the Sun, this procedure cannot be used, since the planet precedes the Sun across the sky. You will therefore need to note the azimuth of the Sun at a given time on the previous day, and then sweep for the planet at the right-ascension time interval *before* the Sun returns to that position. This is more involved, but you will find that, once Venus has been located on a single occasion, it can be re-found surprisingly easily by simply estimating the position of the area in which it lies. The difficulty is to find it the first time!

It should not be assumed that the altitude of the planet will be the same as that of the Sun. It may be as much as 10° to 15° higher or lower in the sky. The almanac will enable you to work this out too, by comparing their declinations, but a rule of thumb may be useful – when Venus is east of the Sun, it tends to be further north than the Sun in the sky in winter and spring, and further south in summer and autumn; the reverse is true when it is west of the Sun. But, with binoculars or a good finder, it does not take long to sweep a few degrees of sky, and you will not need to know its true altitude with any great accuracy in order to locate it, if the sky is favourable. It is always exciting to locate an astronomical object in the daytime sky, and Venus is by far the best candidate for the hunt.

The Superior Planets

Since the Earth's orbit lies 'inside' those of the superior planets, they appear to traverse the Zodiac independently of the Sun. This means that they can appear above the horizon all night, if they happen to lie in the opposite direction to the Sun. This position is known as *opposition*, and it is the most favourable time for observation, since the two worlds are then at their closest. In between successive oppositions, the superior planet will pass through the unobservable *conjunction position*, when it is more or less behind the brilliant solar disc.

Mars

Mean solar distance: 228 million km (the orbit is relatively eccentric, and the distance varies from 207 million km at perihelion to 249 million km at aphelion).

Length of year: 687 days.

Average time between oppositions: 780 days.

Diameter: 6794 km.

Mars is the first of the superior planets. It appears to circle the Zodiac independently of the Sun, and comes to regular oppositions, when it rises at sunset and remains above the horizon all night. This is the time when a superior planet is closest to the Earth and therefore best placed for observation.

When Mars swings near, which it does every two years and two months, there is a special feeling of excitement amongst amateur observers. Even though the *Viking* spacecraft seem to have demonstrated the absence of anything resembling 'life', it is still an intriguing planet, half-Moon, half-Earth. We could not tell this before spacecraft took close-up pictures, since its disc is usually very small, and only the grosser features can be made out. The largest lunar-type crater would be undetectable at the distance of Mars, as is its tremendous volcanic peak Olympus Mons, some 25,000 m high.

Prepare to be disappointed when you turn a telescope to Mars. With luck, a 60-mm refractor and a high magnification will reveal a white spot at one of the poles: a true polar cap of water ice, but no more than a thin surface deposit which sublimes into the tenuous atmosphere in the summer

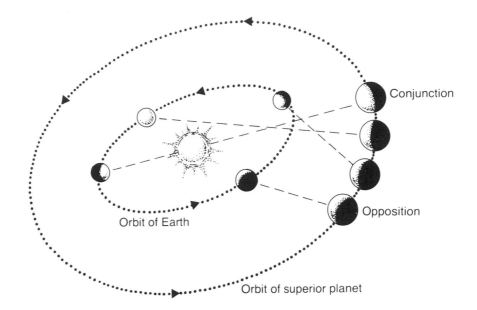

26 *How a superior planet appears to move, as seen from north of the Earth's orbit. The best view is at opposition; the worst at conjunction, when it appears to be near the Sun in the sky. Notice how the planet will appear to move backwards in relation to the Earth after opposition. (Not to scale.)*

Conjunction

Opposition

Orbit of Earth

Orbit of superior planet

months.* If you follow the planet for some weeks, you may find the cap shrinking or growing, according to the season; you may also find the dark areas growing hazy or invisible beneath the tremendous dust-storms which regularly smother huge areas of the planet around the time of perihelion, settling in due course and allowing the permanent features to reappear.

Mars looks warm, because of its reddish colour, but really it is bitterly cold, dropping to below −100°C at night and never rising above about −40°C, even at the equator. Its atmosphere is a hundredth the density of our own. When you have a good view of it, you may notice that the limb of the planet appears brighter than the centre of the disc. You are looking at the sunrise and sunset terminators, where the Sun is very low in the sky, and hoar frost is sparkling on the surface, with perhaps also a frosty haze in the tenuous air.

Oppositions of Mars The way Mars presents itself for inspection from the Earth governs our view of this planet. First, there is a very long gap between oppositions − over two years. This is the longest interval between the favourable presentation of any major planet, and it happens because the rate at which Mars orbits the Sun is not all that much slower than our own. Therefore, after opposition, the Earth only slowly draws ahead of Mars, and it takes over a year for the two planets to come to *conjunction*, when

*The ice does not pass through the liquid (water) state, since the atmospheric pressure on Mars is so low that it instantly evaporates – like solid carbon dioxide on the Earth's surface. Because of the low atmospheric pressure, ice will evaporate at a much lower temperature than on Earth.

they are stationed on opposite sides of the Sun; another year elapses before the subsequent opposition. The situation is very different for the outermost planet, Pluto. It moves so slowly, taking 248 years to orbit the Sun, that the Earth returns to opposition after only a year and $1\frac{1}{2}$ days!

Therefore, oppositions of Mars are rare events, and eagerly anticipated. But there is another problem: since its orbit is so eccentric, it can be much nearer the Earth at some oppositions than at others. If opposition occurs when Mars is at perihelion, its distance will be only 57 million km; at an aphelic opposition it will be as much as 99 million km – almost twice as remote – which means that its disc appears only half the diameter in the sky. Opposition distances follow a cycle, and the following table shows how large it will appear on forthcoming occasions.

Opposition date	Diameter
10 Jul 1986	23.1″ arc
28 Sep 1988	23.7″ arc
27 Nov 1990	17.9″ arc
7 Jan 1993	14.9″ arc
12 Feb 1995	13.8″ arc
17 Mar 1997	14.2″ arc
24 Apr 1999	16.1″ arc

To give you some idea of what these diameters mean, the lunar crater Copernicus appears almost 50″ arc across, so that Mars very rarely appears even half its size. Remember, too, that its apparent size changes perceptibly week by week around the time of opposition, so it is best to try and observe it whenever it is favourably placed in the sky.

Retrograde motion Mars is fun to watch with binoculars or the naked eye, not because you can see its disc, but because it seems to move quite rapidly in front of the stars. Try picking it up in the morning sky, when it rises not long before the Sun (perhaps making a special early-morning effort once a week!), and follow it through towards opposition, when it rises at dusk. If you make drawings showing its position relative to the nearby stars in the constellation in which it lies, you will notice a very curious thing. Some months before (and after) opposition it moves steadily in front of the stars in a west-to-east direction, which reflects its true direction of orbital motion around the Sun. However, about five weeks before the date of opposition it will have come to a halt, and then it will back-track, from east to west. This is *retrograde motion*, caused by the speedy Earth overtaking it, as it were, on the inside, and making the more distant planet appear to move backwards (see Figure 26). This back-tracking effect is greatest at the time of opposition. Some five weeks after opposition it will have come to another stop, and then the planet will continue on its easterly way. It is most interesting to chart this motion and to draw the loopy path of Mars in front of the stars, using an atlas.

Notice, too, how the brightness of Mars changes as it passes through its cycle of visibility (this interval, from its appearance as a 'morning star', through opposition, to its final disappearance as an 'evening star' in the western twilight over a year later, is known as an *apparition*). At a perihelic opposition, its magnitude can reach about −2.5, which is even slightly brighter than Jupiter, and far more brilliant than any star. Mars then shines as a glorious pale ruby object. But near conjunction it is only about magnitude 1. Compare its brightness with that of some stars in the sky. This is not easy, because Mars is reddish and most stars are white, but if you do manage to make the comparison you will find it fascinating to log the change, and relate it to the planet's motion towards and away from the Earth.

The Minor Planets or Asteroids

Between the orbits of Mars and Jupiter, there is a vast gulf of 550 million km. If you look at a scale plan of the major planets' orbits, there seems to be something missing between these two worlds: the regular series of paths is interrupted. This logic was enshrined, in the eighteenth century, in *Bode's Law*, and it led to the discovery of numerous small planetary bodies circling in this gap. Evidently, although a proper planet tried to form here, its particles were disrupted before they could achieve a worthy size (no doubt the fierce gravitational force of Jupiter was partly responsible). The largest fragment is Ceres, 1000 km across. A few dozen are 100 km or more across, but most must be tiny indeed, and undetectable at their great distance.

One minor planet, Vesta, reaches the 6th magnitude at a perihelic opposition, and is theoretically detectable with the naked eye at such times – but these happen only every few years. (Vesta is much smaller than Ceres, but reflects light more efficiently, being made of more metallic material.) Some can be picked up with binoculars, being in the 6th to 9th magnitude range. The difficulty lies in knowing where to look for them, and to have a chance of detecting a minor planet you need to consult an almanac which predicts their positions. Use the *Astronomical Almanac* to locate the four brightest (Ceres, Pallas, Juno and Vesta), marking their nightly positions on a star atlas to act as a guide. They are not likely to be visible in the sky simultaneously, but you will probably find that at least one of them is due to pass through opposition in the next few months.

Now the fun begins! Use the largest-scale star atlas you can find to mark the night-to-night positions as accurately as possible. Draw a line through these points to represent the minor planet's path over the next month or so. Now take binoculars or telescope and locate the field of view for the particular night in question. Somewhere in this field, your minor planet is (or should be) lurking; but which faint speck of light is it? Unless you have an atlas which shows stars down to the faintness of your quarry, you have no instant means of telling which is the intruder; you have to wait for its orbital motion to reveal its identity. So what you have to do is to draw all the 'suspects' accurately in relation to each other, and then re-observe the field on the following night, to find out which one has moved.

This may sound straightforward, but in practice it rarely is! When you return to the field on the next occasion, you may find it hard even to identify the star pattern, perhaps because the night is hazy, or so clear that many other faint stars have come into view to confuse you. Even when you are sure that you are looking in the right place, several of the stars may appear to have moved, due to inaccuracies in your drawing, so you have to make a fresh map and start again. Another possibility, or even probability, is that a succession of cloudy nights allows your quarry to slip away out of the field altogether, and the chase must shift to a new region of sky.

All these obstacles make the final definite identification all the more worthwhile. When the chase is finally successful, you will feel as pleased as you did when you first saw Mercury in the twilight: inspired, you may go on to look at the rest of the four brightest asteroids. They look just like faint stars, but you now know that they are not, and that you have joined the privileged group of people who have knowingly identified some of these strange worlds!

Jupiter

Mean solar distance: 778 million km.

Length of year: 11 years 10 months.

Interval between oppositions: 13 months.

Diameter: 142,700 km (equatorial); 133,200 km (polar).

Jupiter is the first of the *giant planets*, a name which distinguishes Jupiter, Saturn, Uranus and Neptune from the four inner *terrestrial planets*. These two groups have important common characteristics: the terrestrial planets are relatively small, have rocky surfaces, and have relatively thin atmospheres or even none at all (thin, that is, in terms of depth – the atmosphere of Venus is something like a hundred times denser than our own, but relative to the size of the planet it is as thin as an onion skin). The giant planets, on the other hand, are huge balls of gas and frozen chemical compounds. Any truly solid surface must lie far down towards their centres, so that to all intents and purposes we can see only the top of a planet-wide waste of cloud.

Jupiter is a splendid planet to observe. It comes to opposition once every thirteen months, is always brighter than the brightest star (round about magnitude −2.0 at opposition), and is so huge that, despite its remoteness, it shows a tiny disc even when binoculars are used. A very small telescope also reveals its four bright satellites. So it is by far the best planet to observe, no matter how modest your equipment.

Jupiter's satellites As befits the largest planet in the solar system, Jupiter has plenty of moons – at least fourteen are known – although the great majority are so small that their diameters have never been measured, and may be less than 100 km across. They resolve themselves into an inner and an outer group, and four of the inner group are large bodies, two being larger than the Moon. The following table gives details of the four large bodies of the inner group, in the order of increasing distance from the planet.

Satellite	Diameter	Orbital period
Io	3630 km	1 day 18½ hours
Europa	3125 km	3 days 13 hours
Ganymede	5275 km	7 days 4 hours
Callisto	4820 km	16 days 18 hours

These are the satellites discovered by Galileo with his tiny spyglass. In fact, they shine as brightly as the fainter naked-eye stars; but they are so close to the brilliant planet that the eye is dazzled and cannot make them out, unless conditions are particularly favourable. If you have excellent eyesight, you might try hiding Jupiter behind the edge of a chimney or other distant object, and searching the sky as near as you can to it – having first checked with binoculars that there *is* a moon favourably placed on that particular side of the planet. If you make this check using an astronomical telescope, remember that right and left are reversed!

You can see from the table that the times taken by the so-called *Galilean* satellites to orbit Jupiter vary from less than two days to a fortnight. This means that their appearance in the sky undergoes a transformation from night to night. Sometimes you may see two little starlike objects on each side of the planet; at other times all four may be on one side, or a couple may be missing altogether, invisible because they are either directly in front of, or behind, the giant planet's disc. (It is possible for all four to be invisible in this way, but such appearances are excessively rare.) When they are all well visible, you will find that Ganymede appears the brightest of the four, and Callisto the faintest. Although large, Callisto's surface does not reflect light as well as the other moons, and we know from the *Voyager* photographs that they are all very different in their surface configuration. Callisto is dead and stony like the Moon, while Io, in contrast, is still volcanically active, with a sulphurous layer of spewed-out magma blotching its face.

Astronomical almanacs will identify the satellites for you, but with a little experience you can work out which they are for yourself, especially if you track their changing positions on successive evenings. Compare sedate Callisto with scurrying Io, the latter always within about 2′arc of the planet, the former reaching to over 10′. You will notice, too, that they always appear to lie on or near an imaginary line passing through the centre of the planet; this is because the planes of their orbits are very accurately in the plane of Jupiter's equator, which we see more or less edge-on. You are not, therefore, likely to mistake a faint nearby star for a satellite, which is something very easy to do in the case of Saturn.

Jupiter's markings Turn a pair of ×8 binoculars on to Jupiter, and you will see that it is a minute disc rather than a pinpoint star – assuming that the glasses are good ones. A ×20 telescope should show a perceptible disc, looking slightly elliptical (this is caused by Jupiter's rapid rotation, in less than ten hours, which causes its equatorial regions to bulge outwards). With higher powers, cloud markings begin to appear. They may seem only to be faint dark lines running parallel to the equator, but you are looking at bands of ammonia and methane clouds that could swallow the Earth in their depths, and freeze it at their surface temperature of about −150°C. The bright parts of the planet are at a higher altitude than the dark bands, and these features are usually called *zones* and *belts* respectively. They always appear in the same general positions on the planet's surface, and Figure 27 shows the names that they have been given.

Usually, the North and South Equatorial Belts are the easiest to see, and you may well notice them first. The South Temperate Belt is also very often prominent. A good 60-mm refractor should show these belts, and others as well if Jupiter happens to be 'active' – in some years its atmosphere seems to produce more observable detail than in others.

It is interesting to spot the belts and zones, but even more so to see individual cloud details within them. You are unlikely to make much out on the first few occasions, unless you are using a large telescope of good

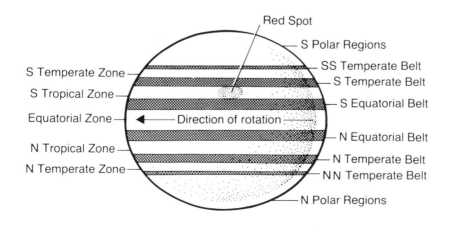

27 The belts and zones of Jupiter, and the position of the Red Spot, as seen through an inverting telescope.

quality, and the air is steady. With practice, however, you will find yourself noticing gaps or wisps in what before seemed to be an even band. Try to keep your eye on one of these details, and follow it. After half an hour, or perhaps even less, it will seem to have moved perceptibly across the disc. This is evidence of Jupiter's rapid rotation, carrying details across before your eyes.

You are bound to want to see the famous Red Spot. This huge elliptical cloud feature, some 40,000 km long, is located in the South Tropical Zone, and makes an indentation in the South Temperate Belt known as the Red Spot Hollow. However, it is not always noticeably red, and recently it has shown little colour in small telescopes, being quite hard to locate. At other times, though, it has been quite vividly reddish, even with apertures of 75 mm. Do not be unduly disappointed if you cannot find it during a 'faint' period, but keep in touch with an active observer, and await a warning of renewed activity, which is bound to occur again some day.

Even without the prominent Spot, however, there is always a feast of detail and interest associated with the solar system's majestic king.

Saturn

Mean solar distance: 1427 million km.

Length of year: $29\frac{1}{2}$ years.

Average interval between oppositions: 1 year and 2 weeks.

Diameter of planet: 120,800 km (equatorial); 109,100 km (polar).

Diameter of ring system: 272,300 km (outer); 149,300 km (inner).

If you own even a small astronomical telescope, and want to inspire your friends with enthusiasm for your hobby, there is one planet above all others that will do this for you: provided, of course, that your telescope is a good one, capable of giving a sharp image with a magnification of × 50 or more. Point it towards Saturn, and wait for the gasp of wonder!

Everyone knows what to expect when they look at Saturn – or do they? It is the planet with the rings, pictured countless times in countless books. But to see it *in the sky*, small though it may appear, is an utterly new and thrilling experience. For the first time, one can believe that those rings really do exist, and one can marvel at the invisible forces that hold them there like some celestial conjuring trick, frozen in time.

So, let Saturn work its magic on you. Treat it as one of the spectacles of the sky that require no insights to be appreciated. With a 60-mm refractor, even an excellent one, you will find it hard to make out much detail, but the clear beauty of Saturn's outline will never fail to evoke its own response.

Saturn's rings Until very recently, astronomers imagined that the ring system of Saturn was unique – certainly within the solar system. However, very faint rings have been detected around both Jupiter and Uranus – although they are far too dim to be seen directly, even with large instruments – making it appear as if ring formation may be a common quirk of giant planets. Evidently, Saturn's rings consist of innumerable blocks or grains of icy material, forced by mutual collisions into an extremely thin sheet in the plane of Saturn's equator. If their orbits crossed and tangled, they would smash into each other, losing or gaining energy, and so either fall catastrophically into Saturn, or be thrown out into space. It is estimated, from spacecraft observations, that the sheet of moonlets is less than a kilometre thick, despite its enormous extent. Therefore, if it is presented edge-on to the Earth, it practically disappears from view.

The presentation does, indeed, vary quite considerably during Saturn's long 'year', the equivalent of $29\frac{1}{2}$ of our own. Figure 28 explains why this is so. The rings lie exactly in the plane of Saturn's equator. If its axis of rotation were vertical, we should always have an edge-on view, with the planet's equator (supposing it were marked in some way) forming a straight line across the centre of the disc. However, the axis is tilted from the vertical at an angle of about 28°. Since the point at which this axis is directed is fixed in space, we see Saturn's globe apparently nod and swing from side to side in the course of its revolution around the Sun. Its ring system does likewise. During one apparition, it may be as fully presented to us as it ever can be; some seven years later, it will be edge-on; after that, the opposite face will

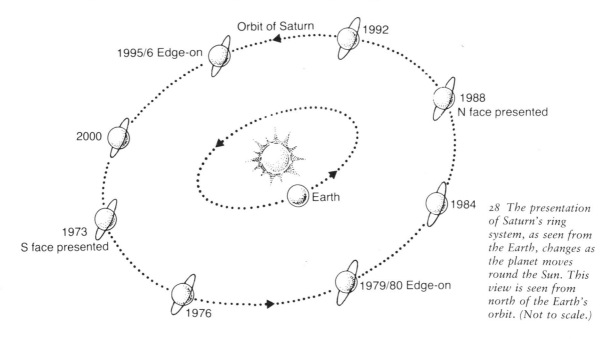

28 *The presentation of Saturn's ring system, as seen from the Earth, changes as the planet moves round the Sun. This view is seen from north of the Earth's orbit. (Not to scale.)*

come into better and better view. At the time of writing (1985) the northern face is presented to us, and the maximum presentation will occur in 1988; the next edge-on phase will happen in 1995.

Examine the rings with the highest power that gives a really sharp view. You will probably see that they are not of similar brightness, and that the outer annulus (Ring A) is somewhat fainter than the inner zone (Ring B). Between these two rings is a gap, known as *Cassini's Division* after its discoverer. This division is about 5000 km wide, and can be detected with an aperture of about 100 mm, under favourable circumstances, as a black line. You will almost certainly need a rather larger aperture to make out the inner Crepe Ring (Ring C), which is a dusky semi-transparent annulus, through which the planet's globe can be seen.

Saturn's satellites About fifteen satellites have been detected, but many of these are very small and faint. The largest is Titan, 5830 km across, and the largest moon in the solar system. This shines at magnitude 8, and is readily visible with binoculars. Next in brightness are Rhea, Tethys, and Dione, in that order. You are not likely to make out any others using a 60-mm refractor, and it is not always easy to identify these, since they are usually scattered indiscriminately around the planet and may be confused with faint stars. Only when their orbits are seen edge-on do they appear to be in a straight line like Jupiter's satellites. Unless you use an almanac to help work out their positions relative to Saturn, you will have to do some detective work, and try to identify them by their night-to-night motion. Here are the details of the four brightest moons, in order of increasing distance from the planet.

Satellite	Diameter	Orbital period	Magnitude
Tethys	1040 km	1 day 21 hours	10.3
Dione	820 km	2 days 18 hours	10.4
Rhea	1580 km	4 days 11 hours	10.0
Titan	5830 km	15 days 23 hours	8.3

You may also care to look for the strange moon Iapetus, which has one normal hemisphere and one very dark one. When the dark hemisphere is turned towards the Earth, its magnitude drops considerably, to about 12, whereas it shines about as brightly as Rhea when the normal hemisphere faces us. It always appears in this brighter mode when near western elongation from Saturn. Being much more remote than Titan, it takes eighty days to complete one revolution. If you manage to identify Iapetus, you should congratulate yourself, for relatively few amateurs have seen more than the four moons listed above.

The Outer Planets

Uranus, with a diameter of about 52,000 km, takes 84 years to circle the Sun once, at an average distance of 2870 million km. Technically it is visible with the naked eye, since its opposition magnitude is about 5.7, but it resembles a star until a magnification of about × 50 is used, when its tiny disc begins to show. Therefore, a good 60-mm refractor will reveal its planetary nature.

To identify Uranus, take its position in the sky from an almanac and plot it on a star chart. Then look for it, preferably using binoculars. It should immediately be obvious as an extra 'star' in the field of view. If in doubt, follow it over a period of nights, when its very slow motion in front of the other stars will become noticeable. At the present time, Uranus lies in the *Milky Way*,* in the constellation Scorpius, near the southernmost part of the Zodiac.

Pluto is currently closer to the Sun than is Neptune, so that the traditional order of the planets is temporarily disturbed. This is a result of its relatively eccentric orbit, which carries it to within 4450 million km of the Sun at perihelion (which occurs in 1989), but out to a distance of 7350 million km at aphelion! Tiny Pluto, whose diameter is only about 3000 km (smaller than our Moon), takes 248 years to complete this huge path. You will need a telescope of 150 to 200 mm in aperture to identify the planet, which appears in the sky as a speck of the 13th magnitude – but the most difficult task is to distinguish it from the numerous nearby faint stars. At present, Pluto is to be found in the constellation Virgo.

Neptune is, at present, the outermost known planet. It is the last of the giants, some 48,000 km across, and takes 165 years to circle the Sun at an average distance of 4497 million km. Given its position in the sky, taken from an almanac, and using an atlas showing stars down to the 8th or 9th magnitude, you should be able to identify it fairly readily, since its magnitude is about 7.8, bright enough to be picked up in binoculars. You may need to watch it over several nights, to confirm its planetary 'crawl' in front of the stars. Even a first-class 60-mm telescope is unlikely to reveal its disc, since it appears only 2.5″ arc across, compared with 3.8″ arc in the case of Uranus. However, in sighting Neptune (at present in the constellation Ophiuchus), you will have the satisfaction of knowing that you have seen to the edge of the Sun's known kingdom.

*The Milky Way, a hazy band of light seen when the night is clear and dark, is our edge-on view of the Galaxy of stars which contains the Sun (see p. 113).

4
Meteors and Comets

When the first manned spacecraft were launched, preparatory to the *Apollo* voyages to the Moon in 1969–73, there was some worry about 'space safety'. Would some tiny flying particles, moving so fast that they could pass through the wall of a spacecraft, pierce numerous tiny and almost undetectable holes, lowering the cabin pressure as the air leaked away? Or would some much larger body collide catastrophically, so that the mission ended in disaster?

These fears were not unfounded, since observers had long ago proved that such particles exist. They are far too tiny to be observed out in space, but the huge bulk of the Earth, whirling round the Sun at a speed of about 30 km per second, ploughs through them continuously. As they hurtle through the atmosphere, they vaporize and disintegrate in a flash of light – a *meteor*, or 'shooting star'.

As far as we know, no manned spacecraft has ever suffered a collision with a meteoric body, although counts of those that have been observed flashing through the atmosphere show that the solar system must contain countless millions of them. They orbit the Sun, just like minute planets: some in swarms, others in apparently random order. They are, presumably, the left-over debris from the epoch when the planets were formed.

The typical meteor is a body smaller than a marble. Anything as big as a large stone will cause a flash of light bright enough to make you look up in amazement – with luck, quickly enough to see the streak of light, like a bright rocket, known as a *fireball*. Conventionally, a fireball is a meteor which appears at least as bright as the planet Venus. Some really large bodies, weighing several kilograms upon entry into the Earth's atmosphere, may contain sufficient material for some to be left upon reaching the Earth's surface. Such an object is known as a *meteorite*. The original orbiting body in space, whether large or small, is called a *meteoroid*.

Observing Meteors

You will probably observe your first few meteors by accident, although it is surprising how few non-astronomers have ever seen one. They certainly are not rare, although, as we shall see, they are more numerous on some nights than on others. To see a meteor, simply select a clear and moonless night, when the stars are shining as brightly as they ever do from your site (it need not literally be moonless, but the Moon should be very low, or a crescent, so that it does not brighten the sky too much). Lie comfortably on an old carpet, or a sun-bed, facing the zenith, and wait. It may seem a long wait before one appears, but there are lots of ways in which you can pass the time – by studying the constellation patterns, looking at the star colours, or tracing the outlines of the Milky Way: all very pleasant pastimes, and valuable for getting to know the night sky better.

Some minutes may pass, and you will be wondering what people see in meteor astronomy, when you will notice a flash out of the corner of your eye, and a streak of light pierces the star patterns. Another wait – and a much brighter object shoots across your field of vision. Then, perhaps, a much longer wait. Will any more be seen? Your question is answered when two flash across the sky within a minute of each other!

The number of meteors you see during any observing session depends mainly upon two things: the clarity of the sky, and the date. The first is obvious enough. If the sky background is bright, either because of town lights or because there is a bright Moon, faint meteors, like faint stars, will be invisible. So it is not much use hoping to see a lot of meteors if you live in the centre of a city, although you will certainly detect the brighter ones and the fireballs just as well as a country observer is likely to do.

If you go to a dark-sky district, the number of meteors you see will increase enormously. This is because their frequency goes up with increasing faintness: there are roughly three times as many meteors of magnitude 5 as of magnitude 4, for example. So it is worthwhile trying, at least once, to spend a couple of hours under the stars, just gazing up at one of those country skies when the constellation patterns are lost in a blaze of fainter stars.

Meteor Showers

We have said that the time of year also influences meteor activity. This is because a large proportion of meteoroids are not random, or *sporadic*, to use the correct term, but belong to a swarm, or *shower*. Astronomers believe that the curious bodies known as comets – crumbly mixtures of dust and ice, which will be looked at later in this chapter – keep shedding their dusty material as they orbit the Sun. Eventually, the entire orbit is polluted with this dust (a term which should not be associated with the fluffy material swept up by the vacuum cleaner, but refers to grains of solid material up to a centimetre or so across), and each particle is a potential meteor. If the orbit of the Earth carries it through, or very near, the orbit of the decayed comet, there will be enhanced meteor activity during the passage. And furthermore, since the crossing-point of the two orbits is fixed, the passage must occur at the same time every year. This is the cause of a meteor shower.

When you observe a meteor shower, you will notice some differences compared with the sporadic meteors of an ordinary non-shower night. In the first place, the meteors tend to be of the same speed and colour – some showers consist of very swift white or blue meteors, while others have slower yellowish meteors, which you can just about follow with your eye as

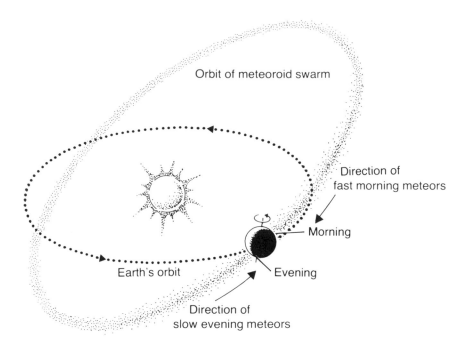

Orbit of meteoroid swarm

Direction of
fast morning meteors

Morning

Earth's orbit

Evening

Direction of
slow evening meteors

29 A meteor shower occurs when the Earth's orbit intersects with that of a meteor swarm. Depending on the direction of the meteors' orbit round the Sun, they will strike the Earth on the 'evening' or 'morning' hemisphere. (Not to scale.)

they cross the sky. This effect of speed does not have anything to do with their velocity around the Sun – it is an effect of their velocity relative to the Earth, or, more correctly, relative to the Earth's atmosphere. If we meet a shower head-on, then the Earth's orbital velocity is added to that of the meteor, so that its destruction is swifter and hotter than if the meteor strikes the Earth side-on.

The velocity of a meteor through the atmosphere also affects its colour. The hotter any body becomes, the whiter is the light it emits. The hottest bodies of all are, in fact, blue-white. Slow meteors are yellowish in tint. This means that you can often distinguish 'shower' meteors from 'sporadic' meteors, seen during the same observing session, simply by their appearance in the sky.

Meteor Radiants

However, there is a much more certain technique for determining the origin of the meteors you see. Imagine that you are observing the sky on a night when the Earth is passing through a swarm of meteoroids. Clearly, they can impact only on one hemisphere of the Earth, since they are all travelling in the same direction. Wherever you stand on this hemisphere, meteoroids are dashing down through the atmosphere above your head. They are travelling through space in parallel paths; but the effect of perspective makes them appear to be fanning out as they approach, so that one may streak away to your right, while another may dart to the left. If you were lucky enough to see several meteors in a shower simultaneously, they would appear to burst apart in the sky from a common centre.

This centre is called the *radiant*. It indicates the direction far out in space from which the meteoroids are approaching the Earth, and it is practically fixed relative to the stars. We only have a good view of the meteors of a particular shower when the radiant is above the horizon. If it never rises very high in the sky, then few meteors will be seen; if it rises near the zenith, then conditions will be excellent, assuming that the sky is clear!

Meteor showers are named after the constellation in which the radiant lies. There are numerous established showers, but most are so feeble that they may not produce more than one or two meteors per hour at best. Obviously, the intense showers are the most spectacular, and there are not very many of these during the year. The following list gives the best showers, with their dates of maximum activity.

Quadrantids	Jan 3–4	Radiant in Boötes; very swift meteors; intense shower
Eta Aquarids	May 5	Swift meteors
Delta Aquarids	Jul 27	A more intense shower than the May meteors
Perseids	Aug 12	Meteors seen over many nights around this date
Orionids	Oct 20	Slow, yellowish meteors
Geminids	Dec 14	Bright and swift

At the time of maximum activity, and assuming that the radiant is well above the horizon in a clear, dark sky, these showers should produce plenty of meteors in the course of a watch. Likely rates range from ten to twenty per hour in the case of the Eta Aquarids to possibly a hundred per hour from the Quadrantids and the Perseids. But, be warned: both these latter showers come to their high maxima very sharply, and in a matter of hours the rates can drop very substantially. If you are unlucky enough to have the radiant low when this maximum occurs, you will see only a few meteors. Conditions change from year to year, and your best course is to consult an astronomical almanac to find out which showers are favourable for observation in a particular year.

Observing a Meteor Shower

The first essential is a good site, and the second essential is comfort. You don't need to be able to see the whole sky, although it is better if you can, but you certainly need to have a clear view of the sky above an altitude of 45° or so, since the majority of meteors are seen in this area. You should also have a view of the sky around the radiant. Seek a spot where no lights shine into your eyes. A public park may be ideal (but beware of prowling policemen, who may wonder what on earth you are up to!), although, as we have seen, a country spot is best of all. Try to get a few friends to join in, even if they don't know anything about astronomy. A group will always see more meteors than an individual will, and you may prefer to be in company anyway. This is one great advantage of joining an astronomical society, which will, or should, organize such watches when a good meteor shower is due.

The second requirement is a comfortable observing position. You need to be both warm and relaxed. Even a mild night can be extremely chilly if you are lying on the ground! So equip yourself with warm clothes, and even a blanket or a sleeping-bag. If you can repose on a portable sun-bed, you will be warmer than if you have to lie on turf. Furthermore, you may be able to adjust the inclination of the head so that you can point yourself at the required altitude. Remember that a meteor watch of less than a couple of hours' duration is hardly worth while, so that comfort and warmth are absolutely essential. Take a flask of strong coffee to keep yourself going!

Assuming that you have checked up that 'this is the night', and are prepared to go out to your observing site, what should you plan to do? You don't have to do anything but enjoy the meteors – but if you want to make a record, then these are things you may care to note down in your book:

1 The time each meteor is seen, to the nearest minute or even half-minute if possible.

2 The brightness of the meteor, estimated in terms of star magnitudes. (Even in an excellent sky, you are not likely to notice a meteor fainter than magnitude 5.)

3 The colour of the meteor.

4 Any unusual features, such as explosions along the path (which usually occur only with bright objects), or a 'train': a luminous afterglow along the track.

5 The length of the path in degrees, from where the meteor was first seen to where it disappeared. You can estimate this by using the outstretched hand, which corresponds to about 20°.

6 If you really want to work hard, try drawing the path on a star map. This is not very easy, but is excellent practice in learning the star patterns. After the observing session, back-tracking these paths will produce a network of intersections in a certain area of the sky around the radiant, showing dramatically the perspective effect referred to earlier. You will probably find that some paths cut across the others, and lead back to nowhere near the radiant. These are sporadic meteors, orbiting the Sun independently.

Don't think, though, that meteor astronomy is something that you can do only a few nights in the year. Some observers carry out watches on every night, waiting patiently for sporadics or members of faint showers. However, even if you do not persevere to this extent, there are times when you may, involuntarily, become an important meteor observer. This is when a brilliant fireball shoots across the sky, distracting you from your other work. Astronomers are interested in these, because a fireball means that the meteoroid is very large, and may, conceivably, survive the fall to the ground. A freshly-fallen meteorite is a rare prize indeed, for very, very few have ever been found. A few good observations of the path can help to locate the possible impact point, and even if nothing is found, they may assist in working out the orbit of the body around the Sun before its fatal collision with our planet.

So, if you see a fireball, forget everything else and try to fix in your mind just two points along its path, relative to the nearby stars. Don't worry too much about the beginning and end points, unless these happen to be easily memorized. Once fixed, make a note of the points you've chosen – and also of the time and any other relevant details. If you see the fireball pass down behind some terrestrial object, then note the exact point at which it disappeared, and also remember the point where you were standing when you made the observation. Subsequent observation by day with a theodolite can turn this latter observation into altitude and azimuth.

Having done this, make sure that you report your sighting to the national amateur astronomical organization, so that it reaches the right people. Really brilliant fireballs often make news, but frustratingly few people make useful observations of them: you, as a serious observer, could help researchers extract some useful results from a mass of vague reports. So, be ready – there could be one tonight!

The Ghostly Comets

We have seen that meteor swarms are almost certainly the debris thrown out by comets. But what *are* comets, and what is their place in the solar system?

A comet is in itself a very small body, just a few kilometres across – which is tiny by planetary standards. But it is not as dense as a planet or an asteroid: it consists of frozen liquids (probably mainly water) and crumbly, gritty material, surrounding a solid centre. In the vacuum of deep space, the icy liquids slowly sublime into a small cloud, so that comets always appear hazy when viewed through a telescope.

Almost all comets have very elongated orbits, and may recede beyond the distance of Jupiter at aphelion, while many approach the Sun closer than Mercury at perihelion. When they are near the Sun, the intense heat causes their outer liquid surface to boil away violently, producing a hazy cloud, or *coma*, thousands of kilometres across. Some of this gaseous material is swept back into a tail pointing away from the Sun, if the comet is particularly active – but not all comets have detectable tails. The coma and tail also contain small dusty particles released from the solid body, or *nucleus*, when the liquids boil off.

There is a popular idea that all comets are bright and spectacular. This is very far from the case, for a dozen or more may be observed every year, but only one or two per decade, on average, achieve real prominence. These particularly bright objects tend to be new discoveries – comets which have never been observed before during recorded history, taking thousands of years to orbit the Sun once. The aphelia of these immensely long-period objects lie far beyond the outermost planets, and may even extend an appreciable way towards the nearest stars.

There is no way of telling when one of these rare visitors will appear. The most famous of the 'regular' comets is Halley's, due back at perihelion in 1986. Unfortunately, it will be a poor spectacle from northern sites, although observers in regions south of the equator should have a better view.

If you want to observe a comet, you must keep in touch with the astronomical news. Most years, one or two are detectable with binoculars or a small telescope, if you know where to look. The monthly magazine *Sky and Telescope* publishes news of recent discoveries, and includes predictions, or *ephemerides*, showing their position in the sky. You may also be able to get news through your local society. Plot the comet's position on a star map, and examine the area carefully, using a low magnification. Sweep for two or three degrees about the marked position, to allow for any inaccuracy in the prediction or your translation. A comet is almost always distinguishable from a star by its hazy appearance, but between themselves they vary considerably: some are mere 'ghosts', detectable only with averted vision, while others have a pronounced sharp bright central condensation. Careful observation over an hour or two will normally reveal motion relative to the stars, proving that the comet is moving through space.

Comets often lie in regions of the sky near the Sun when at their brightest, and twilight can make them difficult to observe. You must, therefore, choose your moment carefully. But this is part of the fun – and few thrills can compare with spotting one of these strange bodies on what may be its first return to the vicinity of the Sun for many thousands of years.

5

Stars and Galaxies

In Chapters 1 and 2, I referred to the way in which the daily rotation of the Earth on its axis, and its annual revolution round the Sun, affect our view of the sky. This chapter examines the background of the night sky more closely: the stars, star clusters, gaseous nebulae, and whole galaxies of stars that you can observe in it, and which will be described in detail in the next chapter.

The Stars Themselves

Appearances are deceptive. The stars in the night sky appear as faint sparks of light assembled in a hemisphere above the horizon. In truth, most of those that you see with the naked eye are more luminous than the Sun; the nearest lies a quarter of a million times further from us than the Sun, and the most remote are hundreds of times more distant than this. They are scattered through the huge volume of space occupied by the Galaxy, an assemblage of young stars and old stars amounting to perhaps 100,000 million individuals. Lost in this throng is our modest Sun. As well as stars, the Galaxy contains immense clouds of 'gust', some luminous, some dark. These are the *nebulae*, and within some nebulae new stars are being formed at this moment. One area of star-formation is the famous Orion Nebula (see page 129), visible with the naked eye.

It is necessary to say a few words about how star brightness is calculated. The system is a very old one, dating back some two thousand years to the time of the Greek astronomer Hipparchus. He called the brightest group of stars *1st magnitude*, the next group *2nd magnitude*, and so on down to the faintest naked-eye stars, which were of the 6th magnitude. When astronomy became more scientific, these grades were subdivided, and star brightnesses are frequently given to an accuracy of a hundredth of a magnitude, although, as far as the ordinary observer is concerned, whole or half magnitudes are good enough. Note that the very brightest stars are now found to fall outside the range as originally devised: the brightest of all, Sirius, in the constellation Canis Major, has a catalogue magnitude of −1.58, and several stars are magnitude 0 or brighter (the smaller the magnitude value, the brighter the star).

A telescope reveals stars much fainter than those of the 6th magnitude. Ordinary binoculars will reach to the 9th, and a large amateur instrument may delve down to the 15th, a class in which there are countless millions of stars.

Each magnitude step represents a brightness ratio of about $2\frac{1}{2}$, and a range of 5 magnitudes corresponds to a brightness difference of exactly 100, while 10 magnitudes mean a brightness difference of 10,000 times. The star

Sirius is about this number of times brighter than the faintest star visible through binoculars or a small telescope.

When you look at the stars of the night sky, you are looking at a very varied collection of objects indeed. Some stars are many thousands of times brighter than the Sun – a case in point being the star Deneb, in the constellation Cygnus (see page 174), which is about 30,000 times as luminous. Yet the star which appears the brightest of all – Sirius, in Canis Major – is only some 15 times as bright as the Sun, and appears brighter than Deneb only because it is so much closer. Therefore, the appearance of a star in the sky is a poor guide to its real luminosity.

If you look closely at the brighter stars of the night sky, you will recognize another difference between them besides their magnitude. This is their *colour*. Most stars appear to be white or whitish, but some are definitely yellow, and a few even appear reddish. It was once thought that a red star and a white star were completely different in composition. But no: apart from minor differences, stars all appear to be remarkably uniform in chemical make-up, consisting principally of hydrogen and helium but with traces of many other well-known elements as well. It is their temperature, not their composition, which affects their colour. Quite simply, a reddish star is cooler than a white star. The hottest known stars have temperatures higher than 30,000°C; the reddest visible may be only about 2500°C.

This is their surface temperature. Deep inside, where nuclear reactions are going on, the temperature is measured in millions of degrees. This is the region where hydrogen is being converted into helium and giving out energy. But the surface temperature is what concerns the observer.

The main sequence, to which the 'normal' stellar population belongs, has already been introduced. I have also indicated that there are exceptional stars which do not lie on this important band. The most important group, as far as the observer is concerned, are the *red giants*. On the main sequence, a red star is small, cool and dim. A red giant, however, is very large, cool and extremely bright. A red giant is an old star, one which has 'evolved', to use the technical term, away from the main sequence and established its own curious make-up. Once hot and white, its outer surface has puffed outwards, enormously increasing its size (a red giant can have a circumference the size of the orbit of Mars). As it puffs out, the surface temperature drops, because it is much further away from the hot energy-giving core. Therefore it reddens; but it is now so huge that it gives out a tremendous amount of radiation. The star Betelgeuse in Orion (page 128) is an example of a red giant, or more correctly of a red *supergiant*, for there are degrees of gianthood!

Another important oddity within the star family is the *supernova*. Some very bright stars, instead of evolving into red giants, simply explode. A supernova explosion is the most violent single event known in Nature, for the destruction of a star in this way releases a burst of energy equal to the output of a whole galaxy of stars. A supernova explosion is effectively over in a few weeks, and the brightest phase lasts only for a few nights. The last supernova observed in our own Galaxy occurred in the year 1604, but it did not appear as brilliant as the supernova of 1572, which shone as brightly as the planet Venus and could be seen in full daylight. Supernovae are sometimes observed in other galaxies.

Star Families

Many stars within the Galaxy appear to be entirely independent. The Sun is one example. But plenty are not. They may have one or more close companions with which they make regular revolutions in the sky. These are known as *binary stars* (if only two are involved), or *multiple stars* if there are three or more. Most of the binaries observable with small telescopes have long revolution periods – measured in decades or centuries – although some very close and hard-to-observe examples go round each other in just a few years. A good example of a fairly easy telescopic binary is the star Castor in Gemini (page 133), the stars of which probably take about $3\frac{1}{2}$ centuries to complete one revolution!

It is worth noting that both binary and multiple star systems are usually included under the general heading of *double stars*. So, too, are the so-called *optical doubles*: stars which are not physically connected, but which happen to lie in almost the same line of sight, and so appear to be close together in the sky. An example of an optical double is the naked-eye pair Alpha (α) Capricorni.

From double stars, to clusters of stars. It seems likely that all stars were born in clusters, because star-formation seems to require a very large, dark cloud of 'gust'. This condenses into a number of nuclei, each of which heats up into a separate star. What happens after star-birth depends upon a number of factors. All the stars and gust-clouds in the Galaxy are travelling round the centre in an elastic merry-go-round way, and interact with each other's gravitational fields as they pass in their separate orbits. This regular tugging tends to pull clusters apart, unless they are very densely packed, and the stars disperse. Undoubtedly, this is why the Sun has no nearby companions now. Therefore, in general, stars escape from their cluster-womb as they age, and most of the clusters that we observe in the sky consist of fairly young stars ('young' meaning up to some hundreds of millions of years old!). A good example is the famous Pleiades group in the constellation Taurus (page 192), which is probably only a few tens of millions of years old. However, some clusters have survived into old age, either because they are very compressed, so that the stars' mutual gravity is strong, or because they happen not to pass near disentangling gravitational fields during their regular circuit of the Galactic empire. Praesepe, in Cancer (page 134), is an example of a fairly old cluster.

An obvious question here is how astronomers know that a cluster is young or old, just by looking at it. The general answer is that a cluster spawns stars of various types, from very hot to very cool. Very hot stars cannot live long, because they burn up their hydrogen so rapidly. Therefore, if a cluster contains these hot stars, it must be younger than one that does not.

There is another type of cluster utterly different from the 'womb' type mentioned so far. This is the *globular* cluster. A globular cluster contains up to hundreds of thousands of stars, and in a photograph looks like a dense swarm of bees. Over a hundred have been discovered in the Galaxy – or more correctly around the Galaxy, because they form a sort of shell around the nucleus, and do not lie in the arms (see p. 113), where all the Sun-type stars are to be found. Globulars are very old indeed, and probably were formed at the same time as the Galaxy itself. Therefore the stars in them are principally old red giants. One or two are bright enough to be seen with the unaided eye, the most conspicuous being Omega (ω) Centauri (page 147), and they are superb sights in a large telescope, although the majority appear as rather faint, hazy patches of light.

Variable Stars

Most of the stars in the sky appear exactly the same from night to night, both in position with respect to each other and in brightness. In truth, their relative positions *are* changing (due to their individual motions through interstellar space), but so very slowly that the general shapes of the constellations remain fixed for thousands of years. Changes of brightness, however, are more readily detectable, and stars which exhibit this fluctuating behaviour are known as *variable stars*. There are two families of variable star: the *extrinsic* and the *intrinsic*.

Extrinsic variables are usually close binary systems. The stars themselves do not vary in brightness, but give the effect of so doing when one star passes in front of the other. The separate stars cannot be made out directly, but their existence can be deduced by analyzing their light. These are known as *eclipsing binaries*. Most eclipsing binaries pass through their cycle in a few hours or days, and some, such as Algol in Perseus, can be followed with the unaided eye quite easily.

A much rarer class of extrinsic variable is the *nebular variables*. These change in brightness because of the motion of dust-clouds in their vicinity. Not many of these are suitable for amateur observation.

The intrinsic variables offer a number of interesting classes. Some are very regular in their performance, and repeat their cycles with great precision. The class known as the *Cepheids*, named after the star Delta (δ) Cephei (page 187), are one such group. Others are not quite so regular; the most famous group are the *long-period variables*, some of which take more than a year to vary from maximum to minimum and back again. A fine example of a long-period variable, or LPV, is the star Mira in Cetus (page 193), which can reach the 2nd magnitude, but sinks almost to the limit of binocular visibility.

There are many other types of extrinsic variable, but special note should be made of the *novae*, close binary systems in which a sudden transfer of hot gas from one star to the other sets off a violent explosion which can lift it from an obscure telescopic object to a bright naked-eye star; and, of course, the extraordinary supernovae, to which we have already referred.

The Nebulae

The Galaxy contains about as much non-stellar mass as stars. A good deal of this is scattered thinly and effectively invisibly throughout space. Elsewhere, however, it masses in denser aggregations many light-years across – the nebulae. Some nebulae are the current birth-place of stars; others seem to be nothing more than huge dormant clouds. Some small nebulae have been expelled from stars; these are known, misleadingly, as *planetary nebulae*, and not many are particularly easy to observe with a small instrument.

Of the large nebulae there are two broad types: the bright and the dark. Bright nebulae shine in the sky as hazy, irregular patches of light. The Orion Nebula has already been mentioned because it is the brightest, but there are plenty of other binocular or telescopic examples. Many of them will be found to contain young, hot stars. In fact, it is these stars which cause the nebulae to glow, because of the radiation which they emit. This radiation is taken up by the atoms in the nebula and re-emitted as visible light. Such nebulae are technically known as *emission nebulae*.

Some other bright nebulae shine simply by reflecting starlight. These *reflection nebulae* tend to be much dimmer. Within the Pleiades star cluster there are faint reflection nebulae, but very few observers have ever managed to detect them, although they can readily be photographed. Reflection nebulae tend to contain mostly dust, whereas it is the gas atoms which produce the light in emission nebulae.

There are also *dark nebulae*. These do not shine because there are no stars within them to produce radiation, so they appear as black outlines against the starry background. Many are visible with the naked eye as irregularities in the Milky Way, but black nebulae also project over many emission nebulae.

Our Galaxy or Star System

If you are standing inside a crowd of people, it may be difficult or impossible to judge how far the crowd spreads, or the shape of the outline. The only way in which you could do so would be to climb on to someone else's shoulders and look around from a higher viewpoint! Our position inside the Galaxy is rather similar – except that we cannot change our viewpoint. Therefore, it has been extremely difficult to develop a picture of what the Galaxy is like.

There are still problems, but the best clue came from observing other galaxies in space and deducing some likely similarities. Many galaxies are flattened, spiral-shaped objects. The flattening of our own can be deduced from the Milky Way effect seen crossing the sky on a clear night. This hazy luminosity is the combined effect of millions of distant stars, merging together in a vast sheet containing the Sun. The spiral nature of the Galaxy was subsequently proved by patient telescopic and photographic work, and the present picture is of a basically two-armed spiral about 100,000 light-years across, with a central mass or hub measuring perhaps 20,000 light-years in diameter. The stars here are old red giants, whereas the stars in the arms are both young and old, and here also we find the nebulae from which fresh stars will be born. A halo of globular clusters completes the picture.

The Sun's position is about half-way from the centre to the edge of the Galaxy. The centre lies in the direction of the constellations Scorpius and Sagittarius, somewhat south of the celestial equator. Here, the night sky is a brilliant mass of stars, clusters and nebulae, and the Milky Way is magnificent. Elsewhere, the plane of the Galaxy can be traced by the Milky Way.

The Galaxy is not an independent star-city, but belongs to a group of perhaps two dozen galaxies, all lying within a diameter of some two million light-years. This is the *Local Group*. The largest member of this group is the famous Andromeda Galaxy, visible with the unaided eye as an elongated haze (see page 181). The other members are considerably smaller, and only one, the galaxy in Triangulum (M 33), is of interest to anyone with a small telescope.

Other Galaxies and Intergalactic Space

With a really big telescope, detectable galaxies outnumber detectable stars. Remember that all the stars we see, whether with the naked eye or with the largest instruments, belong to our own star-system. All the other galaxies of the universe are seen through this foreground scatter of stars. Take a photograph of the night sky in almost any direction, and you will discover myriads of hard, sharp, bright dots (stars in the Galaxy), apparently interspersed with elliptical or wispy objects, some single, others clustered together. These are remote galaxies, most of which are larger and brighter than our own, since at great distances only the exceptionally bright objects can be made out at all. The most distant detectable galaxies are many thousands of millions of light-years away – it is impossible to say exactly how far, since the intergalactic 'yardstick' used to measure these distances is not yet finally calibrated. But it can be said that they are some thousands of times as distant as our near neighbour in the constellation of Andromeda. If you take a photograph, using a big telescope, in some directions in the sky, a single picture can include countless thousands of galaxies!

The nearer galaxies are visible in a small telescope, but it must be admitted straight away that they are not spectacular, appearing as little more than misty specks. Therefore, the interest is in detecting them at all, rather than in the view. Some are spiral, like our own; others are *elliptical*, without any arms; some are shapeless or *irregular*. However, these differences are well seen only on photographs, or with much larger apertures than you are likely to possess.

A Universe on your Doorstep

In the next chapter, I will survey the night sky for the objects described here, and choose the choicest items from the vast storehouse of space. Some are bright, others faint; some are relatively nearby, others are inconceivably remote; some may disappoint, but others will delight.

I have already mentioned the subject of telescopes and observing methods, and some more hints will be given in the subsequent pages. But a few words must be said about star maps. The ones given here are intended to help you make a start, and it will be possible to identify all the listed objects using them; but if you want to be more ambitious and chase more stars and other deep-sky objects, then you may need other maps. One of the best star atlases is *Norton's Star Atlas and Reference Handbook*, published by Gall and Inglis. This shows all the stars in the sky down to the 6th magnitude, as well as many clusters, galaxies, etc. For telescopic work to a fainter level, however, you cannot do better than purchase *Sky Atlas 2000* (Sky Publishing Corporation), which shows stars down to the 8th magnitude as well as numerous interesting objects. With both of these, you will be well equipped to embark on the longest journey anyone on this planet is ever likely to make!

6

Observing the Heavens

The purpose of this chapter is to give you a month-by-month guide to the night sky. It will be usable wherever you are on the Earth's surface, since each month's notes will include two sets of maps, one for observers north of the equator, and another for those in the southern hemisphere. Each set consists of two views, one showing the northern half of the sky and the other showing the southern.

The Monthly Hemisphere Maps

These small-scale maps are intended to help you orientate yourself, and they do not show the constellations in any great detail, simply because to attempt to do so would smother the sky in a confusing mass of stars! In general, stars fainter than the 3rd magnitude are excluded, except where they help to define a constellation more adequately. Increasing dot size indicates the following magnitude groups:

Magnitude 2.6–3.5 •
Magnitude 1.6–2.5 ●
Brighter than 1.5 ●
Key Stars ◉

No attempt is made to label every star, but some of the brightest, which have names of their own, are indicated. These include what I have called Key Stars. Star nomenclature is described in more detail below.

Orientation

These maps are drawn correctly for an observer situated either 45°N or 35°S of the terrestrial equator. These latitudes have been selected because they are near where the majority of amateur astronomers live! However, it is quite possible that you live, for example, in northern Scotland (latitude 58°N), or in Queensland (latitude 15°S), some distance away from the standard latitude. Does this matter? The only difference is that the stars in the northern and southern sky appear at a different altitude above the horizon: the *patterns* of the constellations remain exactly the same, and so once you have identified one or two bright 'marker' stars, the rest can be filled in as easily as if you were in a standard latitude. To help you know what to expect, here is a guide:

Suppose you are in the *northern* hemisphere . . .
If your latitude is *greater* than the standard, then stars in the north will appear higher in the sky by the same number of degrees as the latitude difference, while those in the south will appear lower. If your latitude is smaller, then the opposite is the case.

Suppose you are in the *southern* hemisphere . . .

If your latitude is *greater* than the standard, then stars in the north will appear lower in the sky by the same number of degrees as the latitude difference, while those in the south will appear higher. Again, if your latitude is smaller, then the opposite is the case.

If you do live some distance away from the standard latitude, you may find it helpful to purchase the Philips' planisphere which corresponds most closely to your geographical position.

Map Time

Remembering that the Earth rotates, carrying the celestial sphere round once in a day and a night, the *time* of observation will affect the positions of the stars in the sky. With every hour that passes, the Earth has turned 15° to the east, and the stars appear to have swung 15° to the west. Furthermore, with every fortnight that passes, the Earth has accomplished 15° of its orbit around the Sun. Therefore, if you observe the sky at a certain hour on a certain date, you will have to observe an hour *earlier* if you go out again a fortnight *later* and want the same view! If you go outside a fortnight later at the same time, then the western stars will have sunk lower in the sky, and fresh ones will have risen in the east.

Bearing in mind this 'hour a fortnight' rule, therefore, it follows that a sky view which is correct at, say, 10 pm on March 15 will also represent the sky as it appears at 11 pm on March 1 and 9 pm on March 30.

The small-scale maps in this chapter do show the sky as it appears at these standard times, month by month. It is true that if you live in a high latitude (above about 50° north or south of the equator) the sky at midsummer is not particularly dark at 9 pm – but to have made the times later than this might be inconvenient for people who have to get up in the morning!

Most countries now use Summer Time, which means that the clocks are moved an hour forward, so that the Sun sets an hour later in the evening. Naturally, this means that the stars also reach a given position in the sky an hour later, too, so that instead of representing the sky at, say, 10 pm, the maps now show it as it appears at 11 pm. Never use Summer Time when recording astronomical observations, but keep to standard time (or winter time) for your particular region.

The Close-up Maps

The all-sky guide maps will allow you to locate bright stars and constellations, but most of the interesting objects which you will want to observe are too faint to be shown on them; even if they were, the scale would be too small to be of much help in locating them. Therefore, in order to help you find these objects, larger-scale charts are also included. The areas covered by these charts are shown on the all-sky maps, so that you can locate them. Usually, but not always, these close-up charts show all or part of a single constellation.

These charts show much fainter stars – down to about the 5th magnitude – and the symbols used are as follows:

Magnitude 4.6–5.5 •   Magnitude 1.6–2.5 ●

Magnitude 3.6–4.5 ●   Brighter than 1.6 ●

Magnitude 2.6–3.5 ●   Key Stars ◉

If more than one constellation is included on the chart, the boundary between them is shown by a dotted line. Also, the lines of right ascension and declination (for every whole hour of RA and every 10° of Dec.) are also indicated, helping you to locate the area on a more detailed atlas if you so wish. However, these charts are completely satisfactory for finding the objects listed in this chapter.

Star and Constellation Names

You will notice that the brighter stars are identified by Greek letters, ordinary numerals, or, in some cases, by ordinary letters. The Greek-letter labels are probably the most useful, since they are, very broadly, applied to the naked-eye stars of a constellation in descending order of brightness, from Alpha (α) to Omega (ω) (assuming that the constellation has enough naked-eye stars to need so many letters!). On the maps, the Greek-letter stars are labelled with the appropriate letter. The following alphabet shows the Greek letters as used in the text and on the maps, with their English forms.

Greek Alphabet

alpha α α	iota ι ι	rho ρ ϱ
beta β β	kappa κ $\varkappa$	sigma σ σ
gamma γ γ	lambda λ λ	tau τ τ
delta δ δ	mu μ μ	upsilon υ υ
epsilon ε ε	nu ν ν	phi ϕ φ
zeta ζ ζ	xi ξ ξ	chi χ χ
eta η η	omicron o o	psi ψ ψ
theta θ ϑ	pi π π	omega ω ω

Sometimes, bright stars have been given a name of their own, and this is also indicated. In the constellation of Aquila (the Eagle), the brightest star is called *Altair*, but is also catalogued as *Alpha Aquilae* (in other words, 'Alpha of the Eagle'). However, there is a shorter way of writing this: *Alpha Aql*. Each constellation has a standard three-letter abbreviation, which can save time if the constellation name is a long one (it is easier to write *Beta CVn* than *Beta Canum Venaticorum*, 'Beta of the Hunting Dogs', for example!). So *Aql* is the abbreviation for Aquila, *CVn* is the abbreviation for Canes Venatici, and so on. The following list gives the abbreviations, and the meanings, of all eight-eight constellation names.

The Constellations

Andromeda	And	Andromeda
Antlia	Ant	The Air Pump
Apus	Aps	The Bird of Paradise
Aquarius	Aqr	The Water Bearer
Aquila	Aql	The Eagle
Ara	Ara	The Altar
Aries	Ari	The Ram
Auriga	Aur	The Charioteer
Boötes	Boo	The Herdsman
Caelum	Cae	The Chisel
Camelopardalis	Cam	The Giraffe
Cancer	Cnc	The Crab
Canes Venatici	CVn	The Hunting Dogs
Canis Major	CMa	The Great Dog
Canis Minor	CMi	The Little Dog
Capricornus	Cap	The Sea Goat
Carina	Car	The Keel
Cassiopeia	Cas	Cassiopeia
Centaurus	Cen	The Centaur
Cepheus	Cep	Cepheus
Cetus	Cet	The Whale
Chamaeleon	Cha	The Chamaeleon
Circinus	Cir	The Compasses
Columba	Col	The Dove
Coma Berenices	Com	Berenice's Hair
Corona Australis	CrA	The Southern Crown
Corona Borealis	CrB	The Northern Crown
Corvus	Crv	The Crow
Crater	Crt	The Cup
Crux	Crx	The Cross
Cygnus	Cyg	The Swan
Delphinus	Del	The Dolphin
Dorado	Dor	The Swordfish

Draco	Dra	The Dragon
Equuleus	Equ	The Little Horse
Eridanus	Eri	The River Eridanus
Fornax	For	The Furnace
Gemini	Gem	The Twins
Grus	Gru	The Crane
Hercules	Her	Hercules
Horologium	Hor	The Clock
Hydra	Hya	The Water Snake
Hydrus	Hyi	The (Southern) Water Snake
Indus	Ind	The Indian
Lacerta	Lac	The Lizard
Leo	Leo	The Lion
Leo Minor	LMi	The Little Lion
Lepus	Lep	The Hare
Libra	Lib	The Scales
Lupus	Lup	The Wolf
Lynx	Lyn	The Lynx
Lyra	Lyr	The Lyre
Mensa	Men	The Table
Microscopium	Mic	The Microscope
Monoceros	Mon	The Unicorn
Musca	Mus	The Fly
Norma	Nor	The Square
Octans	Oct	The Octant
Ophiuchus	Oph	The Serpent Bearer
Orion	Ori	Orion
Pavo	Pav	The Peacock
Pegasus	Peg	Pegasus
Perseus	Per	Perseus
Phoenix	Phe	The Phoenix
Pictor	Pic	The Painter
Pisces	Psc	The Fishes
Piscis Austrinus	PsA	The Southern Fish
Puppis	Pup	The Poop
Pyxis	Pyx	The Compass
Reticulum	Ret	The Net
Sagitta	Sge	The Arrow
Sagittarius	Sgr	The Archer
Scorpius	Sco	The Scorpion
Sculptor	Scl	The Sculptor
Scutum	Sct	The Shield
Serpens	Ser	The Serpent
Sextans	Sex	The Sextant
Taurus	Tau	The Bull
Telescopium	Tel	The Telescope
Triangulum	Tri	The Triangle
Triangulum Australe	TrA	The Southern Triangle

Tucana	Tuc	The Toucan
Ursa Major	UMa	The Great Bear
Ursa Minor	UMi	The Little Bear
Vela	Vel	The Sails
Virgo	Vir	The Virgin
Volans	Vol	The Flying Fish
Vulpecula	Vul	The Fox

The Interesting Objects

In addition to mentioning the more important constellations that have come into view, about a dozen individual objects per month will be described in some detail. These could be double or variable stars, nebulae or clusters, or distant galaxies. All will be identified on the close-up charts by the following symbols and identifications:

Double stars These are shown as ordinary stars, since their duplicity was usually not discovered until after they had been catalogued as 'single' stars. For example, the bright naked-eye star Beta Cygni (page 175) is a double star when viewed through a telescope. Note that some doubles are fainter than the bright Greek-lettered stars; these may be identified by a number, such as 5 Puppis (page 135), or by a small letter, such as k Puppis on the same page. These designations refer to different catalogues.

Variable stars These, also, are shown as ordinary stars, since many were catalogued in the ordinary way before their variability was discovered. An example is the famous variable star in Cetus known as Omicron Ceti (page 193). However, some fainter variable stars may have acquired a capital-letter designation. For example, a variable star in Corona Borealis (page 151) is known as R Coronae Borealis.

Star clusters, nebulae, etc. These 'extended' objects have been catalogued by many observers, but two catalogues are sufficient for almost all purposes. The first is the list of 110 objects compiled by the French observer Charles Messier in 1783. These are known as M11, M42, and so on. You will find three Messier objects (M36, M37 and M38) close together in the constellation of Auriga, as shown on page 128.

A much more comprehensive list is the *New General Catalogue* (NGC) of 1888. Thousands of objects are included, many extremely faint. They include almost all the Messier objects, but a Messier number is used in this book, if the object possesses one, in preference to an NGC number. NGC objects are identified on the charts by number only – NGC 1857 in Auriga, near the Messier objects I have just mentioned, is an example.

The symbols used are as follows:

Open star cluster ◌   Planetary nebula -◇-
Globular star cluster ○   Galaxy ◯ ◯
Diffuse nebula ◠ ▢

Against the notes for each object I have included a suffix, as follows:

NE Observable with the naked eye
  B Observable with binoculars
  T Observable with a telescope of about 60 mm aperture

It will be noticed that *double stars* are described in a special condensed way. For example, look on page 128 and find the entry for Omega Aurigae, which runs as follows:

Omega (ω) 5.0, 8.0; 5.8″; 360. White, bluish. (T)

The first two entries (5.0 and 8.0) are the magnitudes of the two stars. The next entry (5.8″) means that they are 5.8″ arc apart in the sky. The final figure, 360, is the *position angle* (PA) of the fainter (mag. 8.0) star with respect to the brighter (mag. 5.0) star. Imagine a clock face, with the bright star at the centre and 12 o'clock marking north (360°), 9 o'clock marking east (90°), 6 o'clock marking south (180°), and 3 o'clock indicating west (270°). Using the position angle information, you can work out in which direction to look for the companion. The second star in Omega Aurigae happens to be directly north of the brighter star, since its position angle is given as 360°.

Finally, the notes indicate that the brighter star is white in colour, while the fainter one looks bluish (although colours do not appear the same to all observers, and you may decide differently!).

Obviously, the 'NE' objects will be seen to far better advantage with optical aid, but this grading will give you some idea of the brightness and scale of the thing you are looking at.

The objects are evenly distributed, as far as possible, between the northern and southern skies, so that an observer in either hemisphere always has something to seek.

Magnitude Test Charts

You will also find two extra circular charts in the pages for January, April, July, and October. These show an area of sky 5° across, centred on an easily-identified bright star – one to the north of the celestial equator, and one to the south. Stars down to the 10th magnitude are depicted, and these will enable you to check the power of a small telescope or binoculars, and help give an idea of what these fainter stars 'look like'.

Some Important Key Stars

Even using the whole-sky maps, it may prove difficult, at first, to identify any of the star patterns. Somehow, a map in a book looks very different to the real sky – for one thing, the scales do not match. Once even one constellation has been identified, things become much easier; but you still have to find this first one!

Possibly you are familiar with the Great Bear (Ursa Major), or with Orion, and can begin from there. If you live in the southern hemisphere, Crux may be a well-known group. But many constellations disappear at certain seasons of the year, when the Sun passes near their vicinity. Therefore, I have compiled a list of the thirteen brightest stars in the sky, all of which are brighter than magnitude 1.0, and each month's notes will give details of where in the sky those *Key Stars* which are visible on that particular occasion are to be found. These directions will be given in terms of azimuth, or compass direction, to the nearest sixteenth of a circle (e.g., west-southwest, south-southeast, etc.), and altitude above the horizon in degrees (remember that from the tip of the thumb to the end of the little finger of an outstretched hand at arm's length is approximately 20°).

Key Stars

Star	Constellation	Mag.	RA		Dec.	
			h	m	°	′
Sirius	Canis Major	−1.58	06	45	−16	43
Canopus	Carina	−0.72	06	24	−52	42
Arcturus	Boötes	−0.04	14	16	+19	11
Vega	Lyra	0.03	18	37	+38	47
Rigil Kentaurus	Centaurus	0.05	14	40	−60	50
Capella	Auriga	0.08	05	17	+46	00
Rigel	Orion	0.12	05	15	−08	12
Procyon	Canis Minor	0.38	07	39	+05	14
Achernar	Eridanus	0.46	01	38	−57	14
Betelgeuse*	Orion	(0.5 var)	05	55	+07	24
(Beta)*	Centaurus	0.61	14	04	−60	22
Altair	Aquila	0.77	19	51	+08	52
Aldebaran	Taurus	0.85	04	36	+16	31
Antares	Scorpius	0.96	16	29	−26	26
Spica	Virgo	0.97	13	25	−11	10

*These are not used as Key Stars, since there is a brighter star in each constellation. They are put in this list for the sake of completeness.

Stars near the horizon are often blocked by local obstructions. Even if the view is clear, they are certain to be dimmed by atmospheric haze, and will therefore appear faint. Therefore, although the hemisphere sky maps show the location of stars towards the horizon, the notes do not mention any Key Star whose altitude is less than about 10° above the horizon.

These directions will be correct for the 'standard' stations of 45°N and 35°S. In general, observers in different latitudes will find altitude affected more than azimuth. However, even if the guidance is a few degrees out, the stars are so well scattered that you are unlikely to make a serious misidentification. The list of Key Stars gives their real name, the constellation in which they lie, their magnitude, and their position on the celestial sphere in terms of right ascension and declination (+ = north of the celestial equator, − = south, which is the common convention). They are given in order of descending brightness.

The January
Night Sky

Northern hemisphere looking north – January

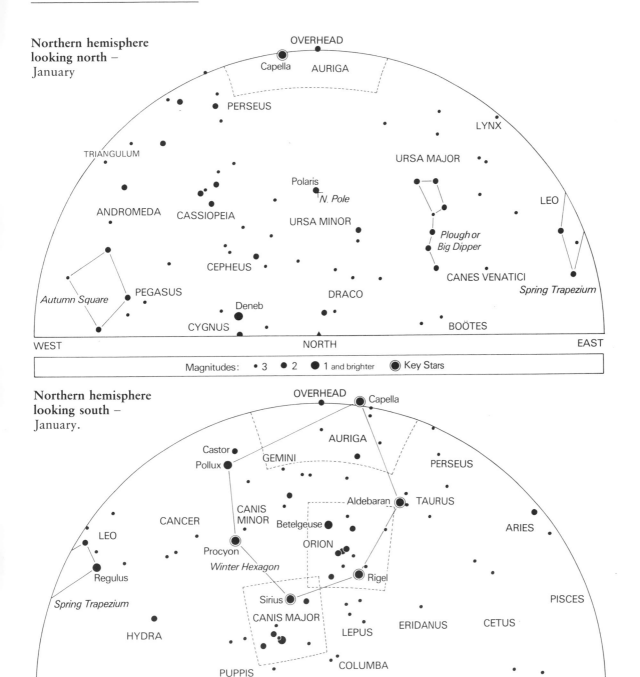

OVERHEAD
Capella
AURIGA
PERSEUS
LYNX
TRIANGULUM
URSA MAJOR
Polaris
LEO
N. Pole
ANDROMEDA
CASSIOPEIA
URSA MINOR
Plough or Big Dipper
CEPHEUS
PEGASUS
CANES VENATICI
Autumn Square
Deneb
DRACO
Spring Trapezium
CYGNUS
BOÖTES
WEST
NORTH
EAST

Magnitudes: • 3 ● 2 ● 1 and brighter ⊙ Key Stars

Northern hemisphere looking south – January.

OVERHEAD
Capella
AURIGA
Castor
PERSEUS
Pollux
GEMINI
CANCER
CANIS MINOR
Aldebaran
TAURUS
LEO
Betelgeuse
ARIES
Procyon
ORION
Regulus
Winter Hexagon
Rigel
Spring Trapezium
Sirius
PISCES
CANIS MAJOR
HYDRA
LEPUS
ERIDANUS
CETUS
PUPPIS
COLUMBA
EAST
SOUTH
WEST

Magnitudes: • 3 ● 2 ● 1 and brighter ⊙ Key Stars

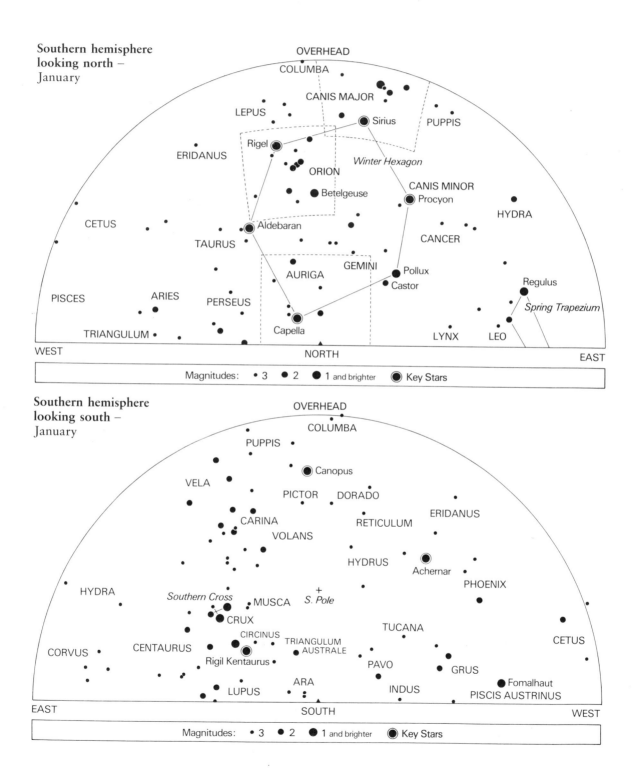

Southern hemisphere looking north – January

OVERHEAD

COLUMBA

CANIS MAJOR

LEPUS

Sirius

PUPPIS

ERIDANUS

Rigel

Winter Hexagon

ORION

CANIS MINOR

Betelgeuse

Procyon

HYDRA

CETUS

Aldebaran

CANCER

TAURUS

GEMINI

Pollux

Regulus

AURIGA

Castor

Spring Trapezium

PISCES

ARIES

PERSEUS

Capella

LYNX

LEO

TRIANGULUM

WEST

NORTH

EAST

Magnitudes: • 3 ● 2 ● 1 and brighter ◉ Key Stars

Southern hemisphere looking south – January

OVERHEAD

COLUMBA

PUPPIS

VELA

Canopus

PICTOR

DORADO

CARINA

ERIDANUS

VOLANS

RETICULUM

HYDRUS

Achernar

PHOENIX

HYDRA

Southern Cross

MUSCA

+ S. Pole

CRUX

TUCANA

CETUS

CIRCINUS

TRIANGULUM AUSTRALE

CORVUS

CENTAURUS

Rigil Kentaurus

PAVO

GRUS

INDUS

Fomalhaut

LUPUS

ARA

PISCIS AUSTRINUS

EAST

SOUTH

WEST

Magnitudes: • 3 ● 2 ● 1 and brighter ◉ Key Stars

Northern Hemisphere
(Latitude 45°)

Looking east: Leo has risen completely above the horizon.

Looking south: Orion is on the meridian, with Auriga above, near the zenith.

Looking west: The Autumn Square is approaching the horizon, with Andromeda above.

Looking north: The brighter stars of Draco are beneath the Pole Star.

Capella, in Auriga, is almost overhead.

Key Stars
Sirius: SSE, altitude 25°.
Capella: overhead.
Rigel: S, altitude 36°.
Procyon: SE, altitude 41°.
Aldebaran: SSW, altitude 58°.

Southern Hemisphere
(Latitude 35°)

Looking east: Most of the sky is rather empty, but distinctive Corvus is rising above the horizon.

Looking north: Orion is on the meridian, with Lepus immediately above.

Looking west: Cetus is setting, and Fomalhaut, in Piscis Austrinus, may be seen low down.

Looking south: There are few bright stars due south.

Canopus, in Carina, is almost overhead.

Magnitude test chart for northern hemisphere – January

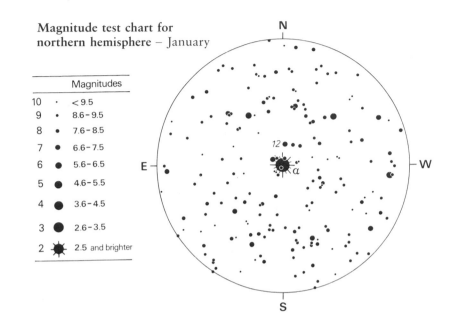

Magnitude test chart for southern hemisphere – January

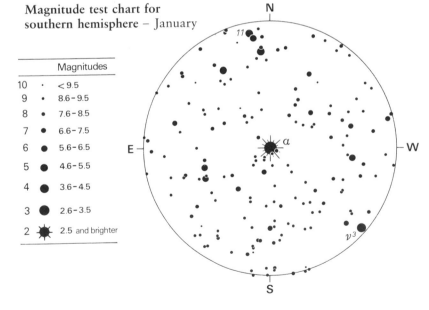

Key Stars
Sirius: NE, altitude 66°.
Canopus: SSE, altitude 69°.
Rigil Kentaurus: SSE, altitude 12°.
Rigel: N, altitude 62°.
Procyon: NE, altitude 40°.
Achernar: SW, altitude 45°.
Aldebaran: NNW, altitude 37°.

Magnitude Test Charts

Northern hemisphere: Alpha (α) Aurigae (Capella) marks the field centre.

Southern hemisphere: Alpha (α) Canis Majoris (Sirius) marks the field centre.

For Northern Observers: Auriga (The Charioteer)

The constellation of Auriga, the Charioteer, lies in a fairly rich region of the Milky Way. Its leading star, Capella, is a yellow star of about the same surface temperature as the Sun, but it is far larger, and therefore more luminous, being about seventy-five times as bright.

Double stars

Theta (θ) 2.6, 7.2; 3.5″; 310. Primary white. This is a delicate and difficult pair. A mag.

10.7 star lies at 54″, PA 300. (T)
Omega (ω) 5.0, 8.0; 5.8″; 360. White, bluish. (T)
14 5.0, 7.2; 14″; 225. White, bluish. (T)
41 5.2, 6.4; 7.8″; 355. Stars both white, beautiful. (T)

Star clusters

M36 Appears smaller than M38, and the stars are more scattered. (B)
M37 A compressed object, containing many faint stars close together. Magnificent with a large instrument. (B)
M38 The third bright Messier cluster in Auriga: contains bright and faint stars. (B)
NGC 1857 A loose cluster of faint stars. (T)

For All Observers: Orion

Most people will be familiar with the magnificent constellation of Orion, the Hunter. Many of the stars are young and very hot, belonging to a loose cluster or *association* in space, lying in a spiral arm of the Galaxy near our own. The red star Betelgeuse, an old supergiant, is only 620 light-years away, compared with a distance of about 1500 light-years for the famous *Orion Nebula*, which lies in the association.

Double stars

Delta (δ) 2.0, 6.8; 53″; 360. Very wide and easy. (B)
Zeta (ζ) 2.0, 5.0; 2.8″; 159. An extremely hard test for a 60-

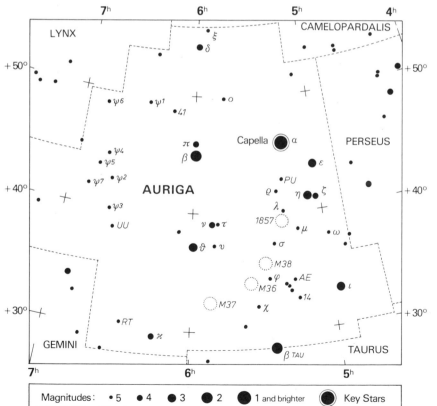

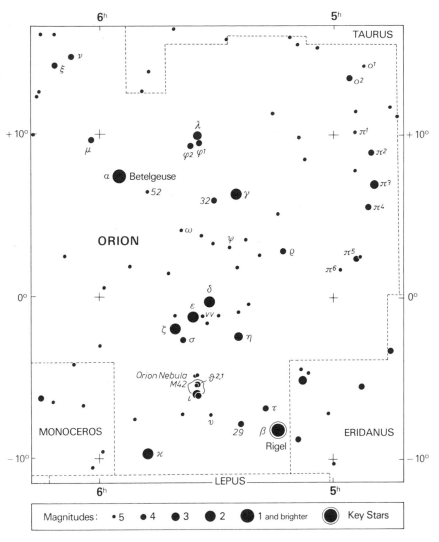

Magnitudes: •5 •4 ●3 ●2 ●1 and brighter ◉ Key Stars

close pair is a good test-object for a 75-mm aperture telescope. (T)

Nebula

M42 The Orion Nebula. With the naked eye, a hazy star; with binoculars, an irregular nebulosity; with a telescope, a convoluted cloud containing some conspicuous stars, and with long wisps of glowing material as well as dark lanes (caused by obscuring nebulae). With a high magnification you may find that the central bright 'star' consists of four close together: the famous *Trapezium* (θ). It is their short-wave radiation which makes the nebula glow. (NE) Look at the star Iota (ι), and see if you can detect a faint haziness around it: this is another cloud of the Great Nebula. (T)

For Southern Observers: Canis Major (The Great Dog)

This fine constellation contains the brightest star in the sky, Sirius, which is only 9 light-years away from the Sun. The Milky Way is quite bright where it passes through this group.

Double stars

Epsilon (ε) Adhara. 1.6, 8.0; 7.7″; 160. A rather difficult pair, for the fainter star is easily lost in the brightness of the primary. (T)

mm telescope, and if you can see the companion you can be well pleased! A high magnification will be necessary, as well as very steady atmospheric conditions. (T)
Iota (ι) 2.9, 7.0; 11″; 141. A delicate telescopic pair, near the Orion Nebula. (T)

Lambda (λ) 4.0, 6.0; 4.2″; 43. Stars white. A fairly high magnification will be needed to separate these stars. (T)
Sigma (σ) A lovely group of four stars close together, of magnitudes 6.0, 7.0, 7.5, and a difficult fourth at 10.0. (T)
52 6.1, 6.1; 1.6″; 212. This very

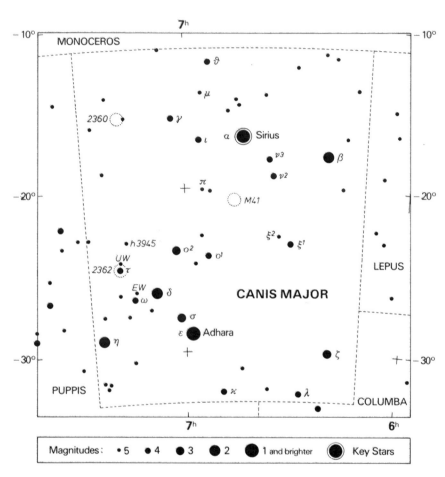

Magnitudes: • 5 • 4 ● 3 ● 2 ● 1 and brighter ◉ Key Stars

Mu (μ) 5.0, 8.0; 3.0″; 339. A very delicate pair. The brighter star is yellowish. (T)
h3945 4.8, 6.8; 26″; 52. An easy pair with any telescope. The primary is distinctly reddish. (T)

Star clusters

M41 A superb bright open cluster, visible with the naked eye as an extensive bright patch. There is a conspicuous yellow star near the centre of the group. (B)

NGC 2360 A cluster of faint stars, of about half the Moon's apparent diameter. (T)
NGC 2362 A much smaller open cluster, easily located since it lies around the star Tau (τ). (T)

The February Night Sky

(Feb 1, 11 pm; Feb 14, 10 pm; Feb 28, 9 pm)

Northern Hemisphere (Latitude 45°)

Looking east: Virgo is rising, with bright Arcturus in Boötes to the west.

Looking south: Canis Minor is on the meridian in mid-sky, with Castor and Pollux, the bright stars of Gemini, up towards the zenith. Very low in the south are the bright stars of Puppis.

Looking west: Aries and Triangulum are fairly high in the sky.

Looking north: Bright Deneb, in Cygnus, may be caught just above the horizon.

In the zenith is Lynx, a very faint group lying between Auriga and Ursa Major.

Key Stars
Sirius: SSW, altitude 27°.
Arcturus: ENE, altitude 6°.
Capella: W, altitude 65°.
Rigel: SW, altitude 28°.
Procyon: S, altitude 49°.
Aldebaran: WSW, altitude 42°.

Southern Hemisphere (Latitude 35°)

Looking east: Virgo is rising, with its bright star, Spica, conspicuous. Above it, the small but very distinctive trapezium of Corvus is well seen.

Northern hemisphere looking north –
February

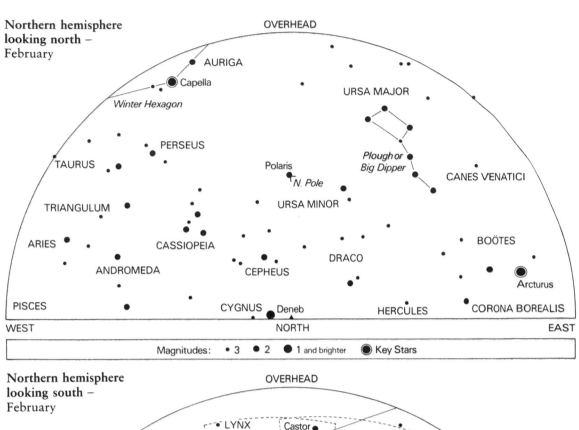

Magnitudes: • 3 ● 2 ⬤ 1 and brighter ⊙ Key Stars

Northern hemisphere looking south –
February

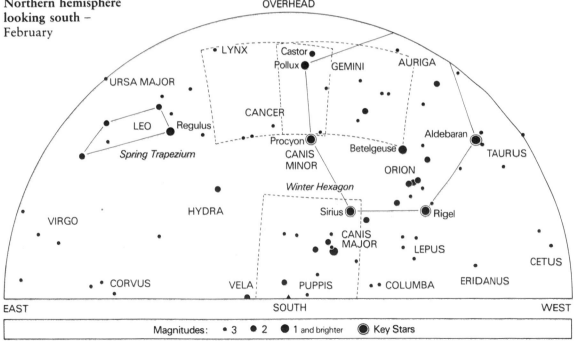

Magnitudes: • 3 ● 2 ⬤ 1 and brighter ⊙ Key Stars

Southern hemisphere looking north – February

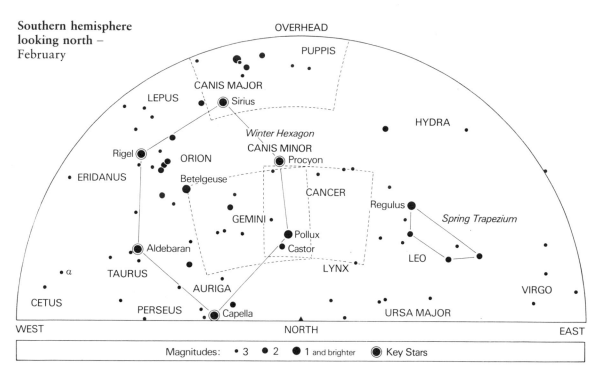

Southern hemisphere looking south – February

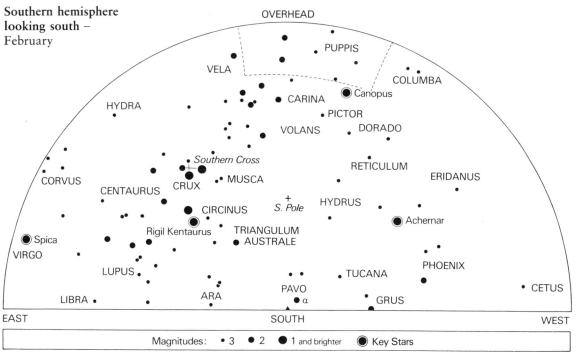

Looking north: The twin stars Castor and Pollux, in Gemini, are distinctive in the lower sky, and Procyon in Canis Minor shines brightly above them.

Looking west: There are few bright stars in this region of the sky. The most conspicuous is Alpha (α) Ceti (Menkar), magnitude 2.6.

Looking south: The low southern sky contains few bright stars, but you may catch Alpha (α) Pavonis (magnitude 1.9) very near the horizon.

The Milky Way, where it runs through Puppis, is in the zenith.

Key Stars
Sirius: NNW, altitude 69°.
Canopus: SW, altitude 68°.
Rigil Kentaurus: SSE, altitude 22°.
Rigel: NW, altitude 48°.
Procyon: N, altitude 49°.
Achernar: SW, altitude 29°.
Aldebaran: NW, altitude 23°.

Magnitude Test Charts

Northern hemisphere: Find Capella (high in the northwest on the northern-facing map), and use the field shown on page 127 (Capella marks the field centre).

Southern hemisphere: Find Sirius (high in the northwest on the northern-facing map), and use the field shown on page 127 (Sirius marks the field centre).

For Northern Observers: Gemini (The Twins)

Gemini, the Twins, occupies a most important position in the sky. It lies on the ecliptic (the apparent path of the Sun round the celestial sphere in the course of the year), so that the Sun, Moon and planets can all pass through its boundaries. Furthermore, it lies at the most northerly part of the ecliptic. This means that the Sun is in Gemini at northern midsummer, and that when a planet lies in this constellation it is most favourably placed for observation from the northern hemisphere, since it passes at its greatest possible altitude above the horizon.

Double stars
Alpha (α) Castor. 2.0, 2.9; 2.5″; 85. A very famous binary star, with a period of about 350 years. A good test for a 60-mm telescope. (T)
Delta (δ) 3.2, 8.3; 6.2″; 220. The bright star is yellowish. (T)
Kappa (κ) 4.0, 8.5; 6.8″; 236. A very pretty pair. The bright star is a lovely rich yellow. (T)
Lambda (λ) 3.2, 10.3; 10″; 33.

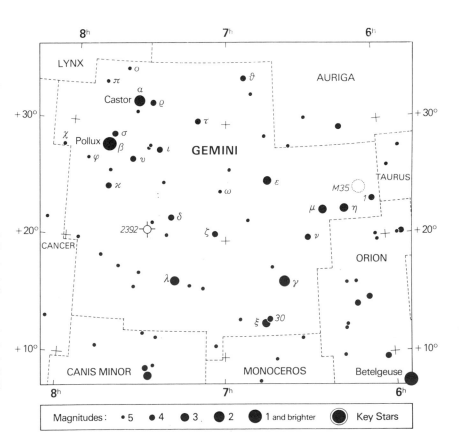

Magnitudes: • 5 • 4 • 3 • 2 ● 1 and brighter ◉ Key Stars

The second star is a difficult object with an aperture of 75 mm. (T)

Variable star

Eta (η) This interesting star changes in brightness very slightly, from about magnitude 3.5 to 4.0. Its period is not regular, but an average cycle takes about eight months. Try comparing it with the star Lambda (λ) (magnitude 3.2) or Iota (ι) (magnitude 3.8) from time to time. (NE)

Star cluster

M35 In clear skies this open cluster can readily be seen with the naked eye as a hazy area, and in binoculars it is very conspicuous. With a telescope, you will see several bright stars in it. The drawback of M35 as a telescopic object is its size. Unless an unusually low magnification is used, its outer regions will be outside the field of view. (B)

Nebula

NGC 2392 This planetary nebula appears as a small faint disc. Without careful scrutiny it may look just like a star. You will have to examine the stars in the telescopic field carefully in order to distinguish it. (T)

For All Observers:
Cancer (The Crab)

This is not a conspicuous group, since the brightest star, Beta (β), is only magnitude 3.6. It also lies on the ecliptic, and

the Sun passes through Cancer round about July.

Double stars

Zeta (ζ) 5.2, 6.1; 5.9″; 79. A lovely yellow pair. The bright star is a close binary, but at the present time the stars are only 0.7″ apart, much too close for a small telescope to resolve. (T) Iota (ι) 4.4, 6.5; 31″; 307. A superb wide and easy pair, yellow and blue. (T)

Star clusters

M44 Praesepe. A very scattered open cluster of bright stars, well seen with bin-

oculars. Try to make it out with the naked eye on a clear night as a ghostly haze. Praesepe means 'Beehive', which is an appropriate name for this swarm of stars. (B)

M67 It is interesting to compare this open cluster with its much brighter neighbour, M44, for it is five times as distant (about 3000 light-years away), and is believed to be a much older object, since it contains many red giant stars, which may have taken a thousand million years or more to evolve from their white-hot youth. (B)

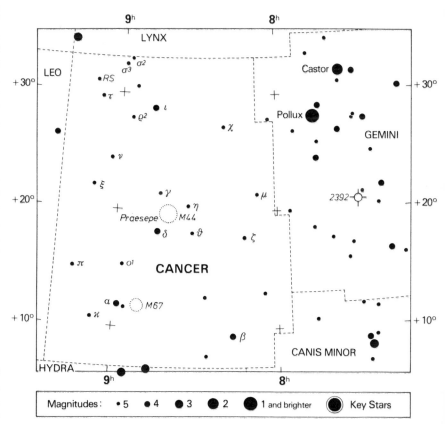

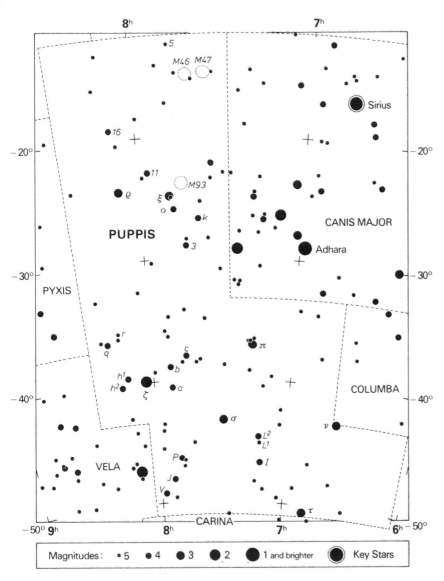

Magnitudes: •5 •4 ●3 ●2 ●1 and brighter ⊙ Key Stars

radiation would fry every living thing on the Earth's surface to a cinder, for it is about 100,000 times the luminosity of our own star!

Double stars

k 4.5, 4.6; 9.9"; 318. Both stars are white, and are almost equal in brightness; you may find it interesting to try to distinguish between them. (T)

r 5.4, 6.2; 4.0"; 190. A fairly close but easy pair. (T)

5 5.3, 7.4; 3.4"; 6. A rather faint and delicate pair. The brighter star is yellowish. (T)

Variable star

L² This is a very red star, and it has been known to change from the 3rd to the 6th magnitude, although like most variable stars it does not perform 'to order', and may fail to show much variation at times. The nearby star I (magnitude 4.5) is a useful comparison (L² is sometimes brighter and sometimes fainter), while at times it may be as bright as Sigma (σ), which has a magnitude of 3.2. (NE)

Star clusters

M46 The stars in this open cluster are not as bright as in M47, but it is still a fine object. (B)

M47 A splendid bright open cluster, visible with the naked eye as a bright patch of light. (NE)

M93 Another bright open cluster. (B)

For Southern Observers:
Puppis (The Poop)

The stars in this constellation are not particularly conspicuous (brightest star Zeta (ζ), magnitude 2.3), but the Milky Way here is magnificent, and you can spend a happy evening just 'sweeping' around with binoculars or telescope. Zeta itself is an interesting star: very young, and hot, with a surface temperature of about 35,000°C. Place Zeta where the Sun is, and the blast of

The March Night Sky

(Mar 1, 11 pm; Mar 15, 10 pm; Mar 31, 9 pm)

Northern hemisphere looking north – March

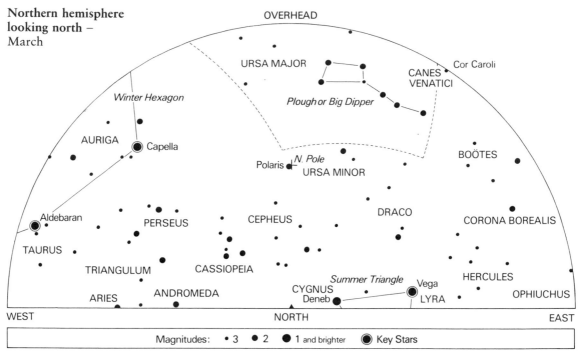

Magnitudes: • 3 ● 2 ● 1 and brighter ◉ Key Stars

Northern hemisphere looking south – March

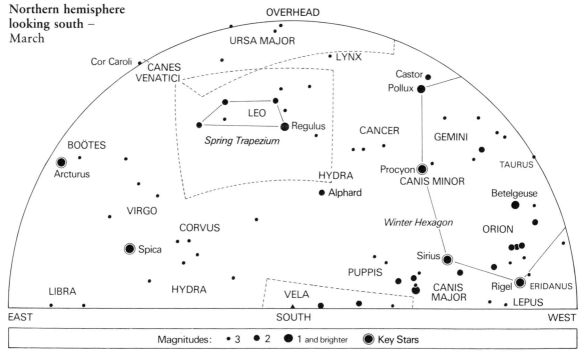

Magnitudes: • 3 ● 2 ● 1 and brighter ◉ Key Stars

Southern hemisphere looking north – March

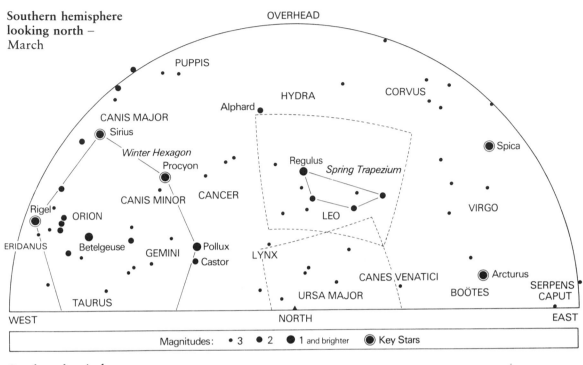

Southern hemisphere looking south – March

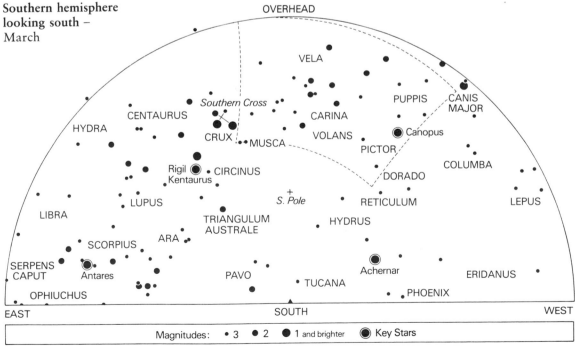

Northern Hemisphere
(Latitude 45°)

Looking east: Boötes is well up, with Arcturus prominent, high in the sky. The small semicircle of Corona Borealis can also be seen.

Looking south: The isolated reddish star Alphard (Alpha Hydrae) is on the meridian, and the head of Leo is also due south now.

Looking west: Taurus is setting, and its bright star, Aldebaran, is at about the same altitude as Arcturus in the east. Orion is also disappearing.

Looking north: The trapezium of Cepheus is underneath the Pole Star.

The 'bowl' of the Plough, Big Dipper, or Great Bear (Ursa Major) is almost overhead.

Key Stars
Sirius: SW, altitude 17°.
Arcturus: E, altitude 25°.
Capella: W, altitude 46°.
Rigel: WSW, altitude 12°.
Procyon: SW, altitude 43°.
Aldebaran: W, altitude 23°.
Spica: ESE, altitude 12°.

Southern Hemisphere
(Latitude 35°)

Looking east: Libra is up, and Scorpius may be caught just above the horizon.

Looking north: The head of Leo, and the bright star Alphard, in Hydra, are on the meridian.

Looking west: Lepus and Orion are approaching the western horizon.

Looking south: The brilliant constellations of Carina and Vela are high in the southern sky.

The zenith contains few bright stars, although the Milky Way runs not far to the west.

Key Stars
Sirius: WNW, altitude 49°.
Canopus: SW, altitude 53°.
Rigil Kentaurus: SE, altitude 35°.
Rigel: W, altitude 26°.
Procyon: NW, altitude 42°.
Achernar: SSW, altitude 16°.
Antares: ESE, altitude 10°.
Spica: ENE, altitude 31°.

Magnitude Test Charts

Northern hemisphere: Find Cor Caroli, the star marked in Canes Venatici (high in the east on both maps), and use the field shown on page 144 (Cor Caroli marks the field centre).

Southern hemisphere: Find Spica (east-northeast on the northern-facing map), and use the field shown on page 144 (Spica marks the field centre).

For Northern Observers: Ursa Major (The Great Bear)

This is a famous constellation, and although it is fairly near the north celestial pole (and therefore its seven bright stars are always above the horizon as seen from north European latitudes), it is at times partly visible from as far south as Australia. From our standard northern latitude, it passes almost overhead on April evenings, but objects at a very high altitude in the sky are inconvenient to observe with a refracting telescope, since a zenith prism must be used to 'bend' the eyepiece into a convenient position. Therefore, the best time to observe this constellation is when it is somewhat to the east or west of the zenith, as now. The name *Ursa Major* means literally 'The Greater Bear', to distinguish it from 'The Lesser Bear' or *Ursa Minor*, which contains the Pole Star (Alpha). However, most people call it simply the Great Bear (and Ursa Minor the Little Bear), or the Plough (which it resembles more nearly). In the USA it is usually referred to as the Big Dipper. It is worth mentioning that the whole constellation extends over much more sky than that covered by the seven bright stars.

If you have identified Ursa Major, and want to find the Pole Star – which will tell you in which direction north lies – then use the stars Alpha (α) and

Beta (β), as shown in Figure 9, to act as pointers to the Pole.

Double stars

Zeta (ζ) 2.1, 4.2; 14″; 150. This double star, Mizar, forms a naked-eye double with the 5th-magnitude star Alcor, 11′ arc away. If you cannot see Alcor with the naked eye, it may be that your sky conditions are too bright (through town lights, probably), or that the problem lies with your

eyesight. Mizar itself is easily resolved with a small telescope. (NE/T)

Nu (ν) 3.7, 10.1; 7.2″; 147. Note the yellow tint to the primary, which is worth looking at even if the difficult companion is invisible. (T)

Xi (ξ) 4.3, 4.8; 2.5″; 94. An extremely difficult test object. Use the highest magnification available, and wait for a night of steady definition, or you will have no chance. (T)

23 3.8, 9.0; 23″; 271. This is a teasing object for a small telescope, and you should be pleased if you succeed in detecting the companion with a 60-mm aperture instrument. The bright star is white. (T)

57 5.2, 8.2; 5.5″; 350. A rather faint but very neat pair. (T)

Nebula

M97 The *Owl Nebula*, a very elusive object. To see this, you must first of all select a site with a truly dark sky, so that the stars appear to shine out against a velvet background. If you live in a town or city, this means a trip out into the country, armed both with binoculars and a small telescope. Next, get thoroughly dark-adapted. Finally, put a very low magnification on the telescope, identify the spot where the nebula should be, and *very slowly* sweep the telescope over the area, deliberately beginning a little above the place where you think it is, and working down with short overlapping swings, like someone pasting wallpaper. In this way you will know that the object must have passed through the field of view in one of the sweeps. If you fail to detect it, try again. You may be surprised at its size, for it appears like a ghostly moon, excessively faint and very large. You can also try for it with binoculars; I have seen it with binoculars quite easily, under favourable conditions. (B/T)

Galaxies

M81 and M82 These are large and important galaxies, lying far beyond the Local Cluster (of which our own Galaxy is the second-largest member), but still relatively very near compared with most other known galaxies, being 'only' 8 million light-years away. This means that the light entering your eye tonight has just come to the end of an uninterrupted journey through space which began long before recognizable humanoids existed on this planet!

Both are visible with binoculars if conditions are good, and they should be readily found with a 60-mm telescope if a low magnification is used. They lie within a degree of each other, so that their different appearance can readily be compared. M81 appears as an elliptical haze with a fairly bright nucleus: it is a spiral galaxy, seen from an almost edge-on direction. Its companion, M82, is an irregular galaxy, apparently suffering an internal explosion, for photographs show material erupting out of it. Visually, it appears as a slightly curved streak of light. (B/T)

For All Observers: Leo (The Lion)

This constellation is one of the very few to look anything like its namesake. It contains a large number of galaxies, although very few are bright enough to be seen in a small instrument. These galaxies also overflow into the area covered by its neighbour Virgo (see April notes).

Double stars

Gamma (γ) 2.2, 3.5; 4.3″; 124. A very slow binary pair, a complete revolution taking about six centuries. The stars are yellow or golden in tint. Close, but quite easy even with a small instrument. (T)

54 5.0, 7.0; 6.3″; 110. An attractive pair, looking white and blue. (T)

Galaxies

The chart shows the positions of some of the brighter galaxies in the Leo-Virgo area. If you want to see these, it is advisable to choose a very dark night, get well dark-adapted, use a magnification of about ×50, and set to work sweeping over the areas indicated. Start off with the brightest: M65 and M66 in Leo. These appear as elongated hazy patches about half a degree apart, and form a true galactic pair in space, lying about 29 million light-years away. Reflect on the immense distance of these faint smudges: each one contains thousands of millions of stars, many of which are more luminous than the Sun! Al-

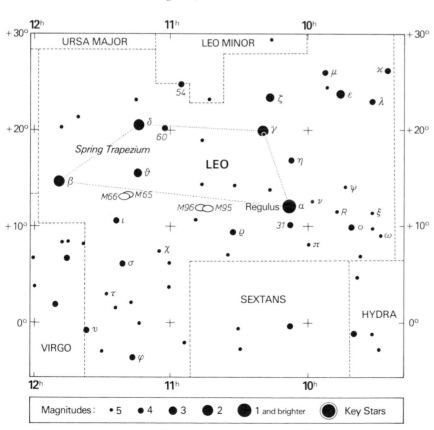

Magnitudes: • 5 • 4 ● 3 ● 2 ● 1 and brighter ◉ Key Stars

though the sight itself is not spectacular, the reflection upon what it *means* should certainly have a stimulating effect. (T)

For Southern Observers: Vela (The Sails) and Carina (The Keel)

These bright constellations lie in a superb region of the Milky Way, and there is much to look at. There was once a huge constellation known in classical times as *Argo* (the Ship), which has since been divided into these two groups, together with Puppis (the Poop). Carina contains Canopus, the second-brightest star in the sky. This is a blazing giant star, about 50,000 times as luminous as the Sun, and over a thousand light-years away.

Double stars
(Carina)
Upsilon (υ) 3.2, 6.0; 5.0″; 128. The brighter star is white. (T)
(Vela)
Gamma (γ) 1.8, 4.2; 42″; 180. A bright and easy pair. The primary is one of the hottest stars known – surface temperature perhaps 50,000°C. The companion may just be visible with powerful binoculars. (B/T)
h 4188 5.5, 6.5; 3.0″; 287. The brighter star is white. (T)
h 4220 5.5, 6.0; 2.1″; 210. A severe test for a 60-mm telescope. (T)

Star clusters
(Carina)
NGC 2516 A brilliant open cluster, visible with the naked eye. (NE)
NGC 3114 A large, scattered open cluster of mainly faint stars. (T)
NGC 3532 A bright open cluster lying in the Milky Way, near Eta (η). (B)
(Vela)
NGC 2547 This bright open cluster can be seen with the naked eye. (NE)

Nebula
NGC 3372 The *Keyhole Nebula* in Carina. An amazing object, visible with the naked eye. Intervening dark nebulae give it an irregular appearance. A remarkable star, Eta (η), blazed up in 1843 from the centre of the nebula, and reached magnitude −1 (almost as bright as Sirius). It is now only visible with binoculars, but may flare up again at any time. (NE/B)

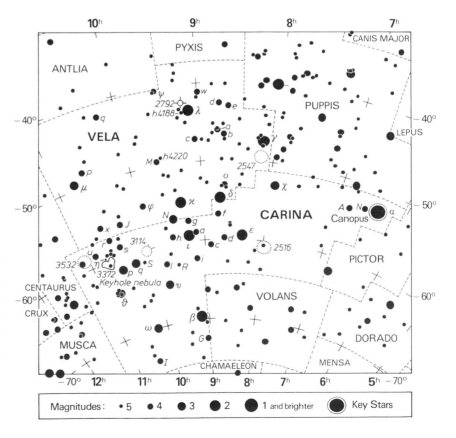

Magnitudes: • 5 ● 4 ● 3 ● 2 ● 1 and brighter ⦿ Key Stars

The April Night Sky

(Apr 1, 11 pm; Apr 15, 10 pm; Apr 30, 9 pm)

Northern hemisphere looking north – April

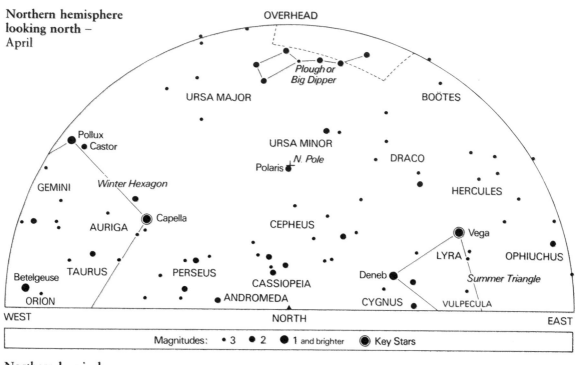

Northern hemisphere looking south – April

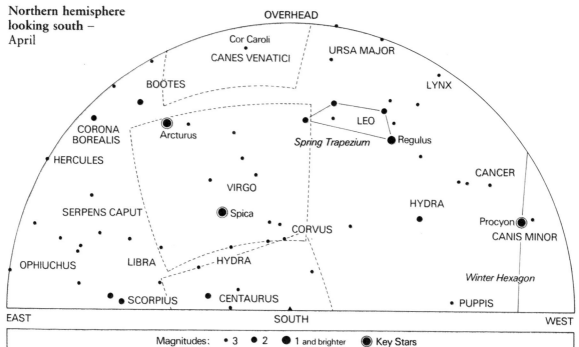

Southern hemisphere looking north – April

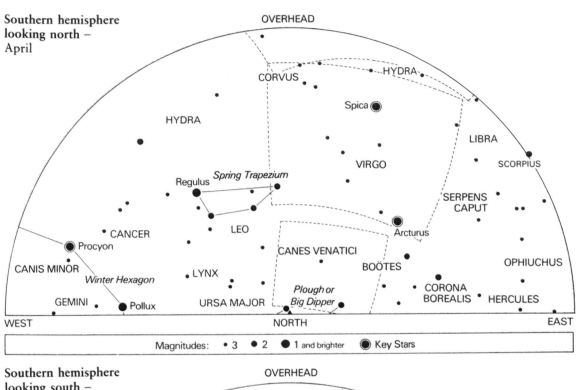

Magnitudes: • 3 ● 2 ● 1 and brighter ⊚ Key Stars

Southern hemisphere looking south – April

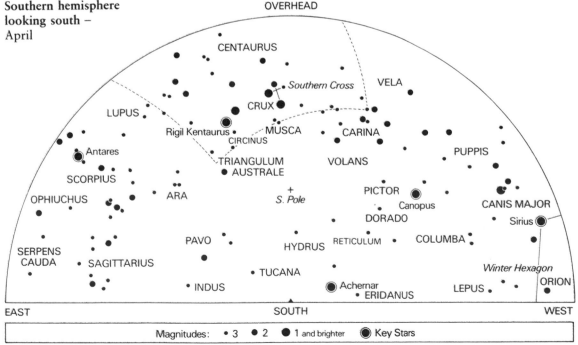

Magnitudes: • 3 ● 2 ● 1 and brighter ⊚ Key Stars

Northern Hemisphere
(Latitude 45°)

Looking east: Hercules and part of Ophiuchus have risen, and the bright star Vega, in Lyra, is very conspicuous.

Looking south: Leo is just past the meridian, and the distinctive group of Corvus can also be seen in the lower part of the sky.

Looking west: Only the upper part of Orion may now be seen above the horizon, with Gemini vertically above it.

Looking north: Cassiopeia is almost underneath the Pole Star.

The Great Bear (Ursa Major) is now in the zenith.

Key Stars
Arcturus: SE, altitude 47°.
Vega: NE, altitude 16°.
Capella: NW, altitude 28°.
Procyon: WSW, altitude 26°.
Spica: SE, altitude 28°.

Southern Hemisphere
(Latitude 35°)

Looking east: Scorpius is now well up, together with most of Ophiuchus and part of Sagittarius.

Looking north: Leo and Virgo occupy this part of the sky.

Looking west: Canis Major and Lepus are setting, but still prominent.

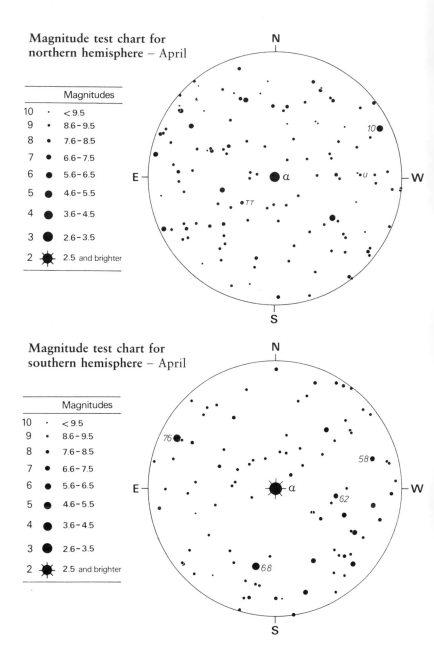

Magnitude test chart for northern hemisphere – April

Magnitudes	
10	< 9.5
9	8.6 – 9.5
8	7.6 – 8.5
7	6.6 – 7.5
6	5.6 – 6.5
5	4.6 – 5.5
4	3.6 – 4.5
3	2.6 – 3.5
2	2.5 and brighter

Magnitude test chart for southern hemisphere – April

Magnitudes	
10	< 9.5
9	8.6 – 9.5
8	7.6 – 8.5
7	6.6 – 7.5
6	5.6 – 6.5
5	4.6 – 5.5
4	3.6 – 4.5
3	2.6 – 3.5
2	2.5 and brighter

Looking south: Crux is high in the sky. You may see Achernar, in Eridanus, not far above the horizon.

The star-poor constellation of Hydra is in the zenith, bordered by the very rich Milky Way areas from Centaurus to Carina.

Key Stars
Sirius: W, altitude 24°.
Canopus: SW, altitude 35°.
Arcturus: NE, altitude 23°.
Rigil Kentaurus: SE, altitude 50°.
Procyon: WNW, altitude 22°.
Antares: E, altitude 26°.
Spica: NE, altitude 54°.

Magnitude Test Charts

Northern hemisphere: Alpha (α) Canum Venaticorum (Cor Caroli) marks the field centre.

Southern hemisphere: Alpha (α) Virginis (Spica) marks the field centre.

For Northern Observers:
Canes Venatici (The Hunting Dogs)

This constellation contains only one prominent naked-eye star, Cor Caroli, which is situated near the focus of the curve of stars Delta (δ) to Eta (η) Ursae Majoris. This is a well-known double star. There are also some deep-sky objects within the ill-marked boundary of this small group.

Double stars
Alpha (α) 3.2, 5.7; 20″; 228. The primary has a yellowish tint. A very easy pair. (T)

25 5.1, 7.1; 1.7″; 105. A close binary pair. (T)

Star cluster
M3 This bright globular cluster can be seen as a hazy spot using binoculars, but it appears very small. A small telescope shows a bright nebulosity of unresolved stars. (B/T)

Galaxies
M51 One of the most famous galaxies in the sky – the *Whirlpool*. This spiral is seen in plan view, and large-aperture instruments do indeed reveal traces of the spiral arms – in a small telescope it appears quite large, but featureless. The most interesting feature is the bright secondary nucleus, which makes it appear as a double nebulosity. (T)
M63 A very elongated nebulosity with a brighter centre. This is a galaxy seen from near the 'edge-on' position. (T)
M 106 This appears as a faint nebulosity with a brighter condensation offset from the centre. (T)

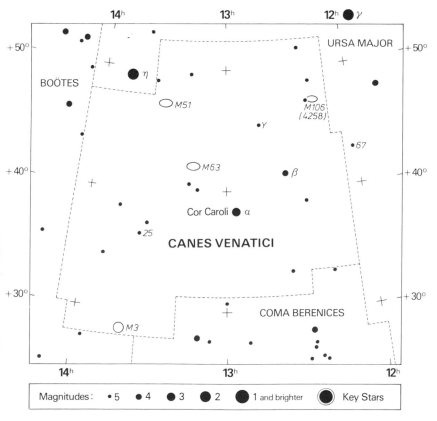

Magnitudes: • 5　● 4　● 3　● 2　⬤ 1 and brighter　◯ Key Stars

For All Observers:
Virgo (The Virgin)

Like its neighbour, Leo (see notes for March), this distinctive, open constellation contains a fine binary star which occupies a place on every amateur astronomer's observing list; strangely enough, too, in both Leo and Virgo the star is Gamma (γ)! However, the main reason for linking these two constellations together is that they contain numerous faint galaxies some tens of millions of light-years away from the Earth. These galaxies do, in fact, form an immense cluster in space. Unfortunately, even the brightest are rather dim objects in a small telescope, but it is certainly worth having a go at finding them.

Double stars

Gamma (γ) 3.5, 3.5; 3.6″; 293. A glorious pair of white stars, with a period of 172 years. They now appear to be approaching each other, and by the year 2000 they will be too close to be divisible with a 60-mm telescope. (T)

Theta (θ) 4.0, 9.0; 29″; 198. This delicate pair may be beyond the range of a 60-mm telescope, but is worth attempting. A very high magnification is not necessary, for the stars are fairly wide apart. (T)

Phi (ϕ) 5.2, 9.7; 4.7″; 110. An unequal pair, with a rather dim companion. An excellent test object for a telescope of 75-mm aperture, and you may congratulate yourself if you can see the companion clearly! (T)

Galaxies

The chart given in the March notes shows the position of some of the brighter galaxies in the Leo region nearby. The brightest of the Virgo objects is probably M49, which can be picked up with binoculars as an obvious haze lying between two 6th-magnitude stars. (B/T) M59 and M60 lie within the same low-power field, M60 being the brighter of the two. (T)

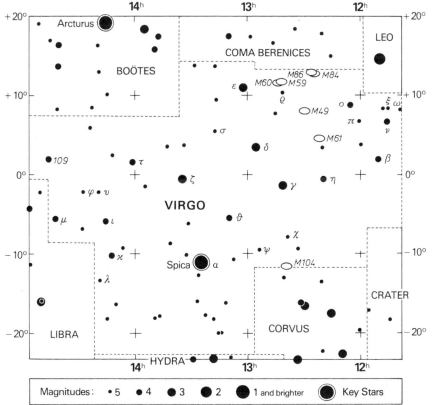

For Southern Observers:
Centaurus (The Centaur) and Crux (The Cross)

Centaurus is easily located, since it contains the two bright stars Alpha (α) and Beta (β), embedded in the Milky Way. Alpha (Rigil Kentaurus) is a magnificent binary star, and is the nearest known star-system to the Sun. Crux can readily be identified by its compact diamond shape, although the westernmost star of the four is rather fainter than the others. Crux is the smallest constellation in the whole sky, but one of the most famous – it is usually referred to as the Southern Cross.

Double stars
(Centaurus)
Alpha (α) 0.0, 1.2; 22″; 212. Very easy: a glorious pair, divisible with almost any telescope. (T)
Gamma (γ) 2.9, 2.9; 1.5″; 356. Superb, but beyond the reach of a 60-mm refractor, although it is possible that the 'star' may appear elongated. This is included in case you have the chance of using a rather larger instrument – perhaps a 100-mm telescope. (T)
D 5.3, 6.5; 2.9″; 245. An interesting test object for a 60-mm telescope. You will have to use the highest available magnification to have much chance of separating the components. (T)
k 4.5, 5.9; 7.6″; 110. Fairly wide and easy. The primary is white. (T)
(Crux)
Alpha (α) 1.6, 2.1; 4.7″; 119. A magnificent pair of white stars, with a magnitude 4.9 star some 90″ away. This star is sometimes known as Acrux. (T)

Iota (ι) 4.7, 7.8; 26″; 25. A small, easy pair. The brighter star is yellowish. (T)
Mu (μ) 4.0, 5.2; 35″; 17. An attractive white pair of stars. (T)

Star clusters
(Centaurus)
NGC 3766 A bright cluster in the Milky Way, near Lambda (λ). (B)
NGC 5139 A superb globular cluster, the brightest in the sky. It appears as a hazy star to the naked eye, and has, in fact, been catalogued as a star – Omega (ω). In binoculars it is a hazy disc of light, and individual stars can be spotted in it using a small telescope. (B)
NGC 5460 A bright open cluster. (B)
(Crux)
NGC 4755 Usually known as the *Jewel Box*, because of the varied tints of its members, or as Kappa (κ). A famous, brilliant cluster, lying very near the prominent dark nebula in Crux known as the *Coalsack*. (B)

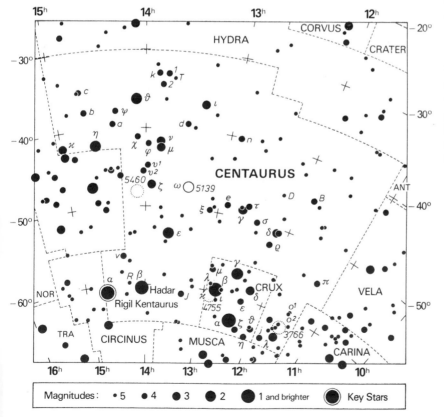

| Magnitudes: | • 5 | • 4 | ● 3 | ● 2 | ● 1 and brighter | ⊙ Key Stars |

The May Night Sky

(May 1, 11 pm; May 15, 10 pm; May 31, 9 pm)

Northern hemisphere looking north – May

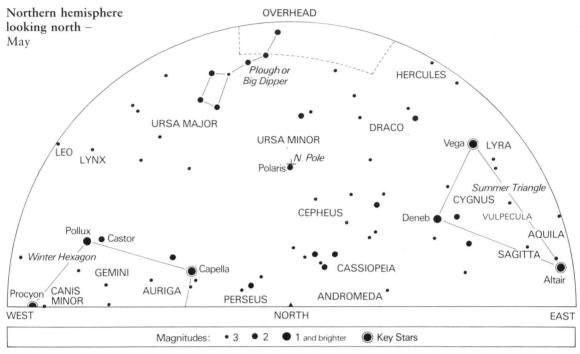

Magnitudes: • 3 ● 2 ● 1 and brighter ◉ Key Stars

Northern hemisphere looking south – May

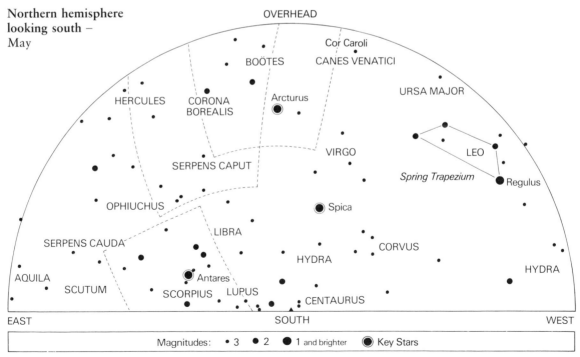

Magnitudes: • 3 ● 2 ● 1 and brighter ◉ Key Stars

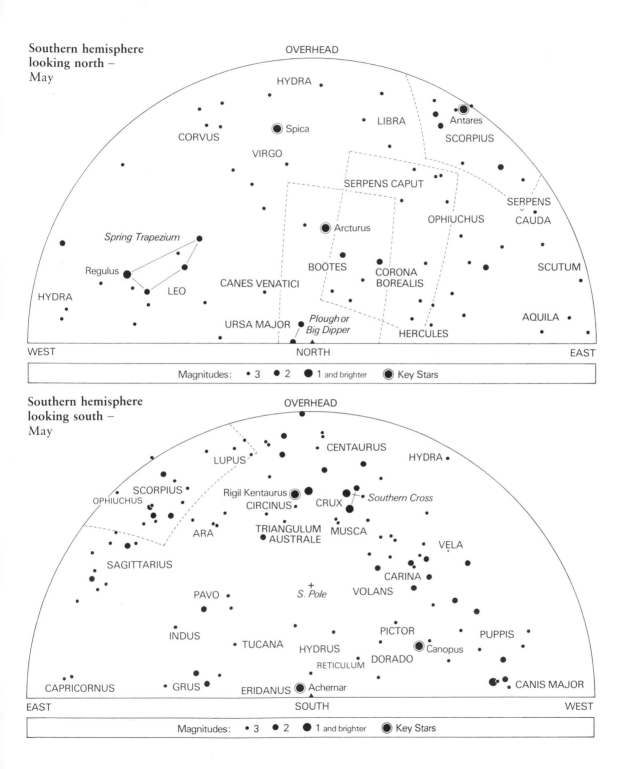

Southern hemisphere looking north – May

OVERHEAD

HYDRA

LIBRA

Antares

SCORPIUS

CORVUS

Spica

VIRGO

SERPENS CAPUT

SERPENS CAUDA

OPHIUCHUS

Spring Trapezium

Arcturus

SCUTUM

Regulus

BOÖTES

CORONA BOREALIS

LEO

CANES VENATICI

HYDRA

URSA MAJOR

Plough or Big Dipper

HERCULES

AQUILA

WEST

NORTH

EAST

Magnitudes: • 3 ● 2 ● 1 and brighter ◉ Key Stars

Southern hemisphere looking south – May

OVERHEAD

CENTAURUS

HYDRA

LUPUS

SCORPIUS

Rigil Kentaurus

CIRCINUS

CRUX

Southern Cross

OPHIUCHUS

MUSCA

ARA

TRIANGULUM AUSTRALE

VELA

SAGITTARIUS

CARINA

PAVO

S. Pole

VOLANS

INDUS

PICTOR

PUPPIS

TUCANA

HYDRUS

Canopus

DORADO

RETICULUM

CAPRICORNUS

GRUS

ERIDANUS

Achernar

CANIS MAJOR

EAST

SOUTH

WEST

Magnitudes: • 3 ● 2 ● 1 and brighter ◉ Key Stars

Northern Hemisphere
(Latitude 45°)

Looking east: The fine Milky Way region from Cygnus to Serpens is now rising, and bright Altair, in Aquila, may be seen low down, with the small constellation of Sagitta above.

Looking south: Virgo, with its distinctive leading star Spica, is on the meridian. Higher up, Arcturus in Boötes is almost due south.

Looking west: Procyon, in Canis Minor, is now very low, and to its right the constellation of Gemini is also setting.

Looking north: The bright constellations of Cassiopeia and Perseus are low in the sky, beneath the Pole Star.

The bright star Cor Caroli (Alpha (α) in Canes Venatici) is almost overhead.

Key Stars
Arcturus: SSE, altitude 62°.
Vega: ENE, altitude 34°.
Capella: NW, altitude 13°.
Antares: SE, altitude 8°.
Spica: S, altitude 33°.

Southern Hemisphere
(Latitude 35°)

Looking east: Sagittarius is now distinctive, well above the horizon.

Looking north: Arcturus, in Boötes, and Spica, in Virgo, dominate the northern sky.

Looking west: The last stars of Canis Major are now setting, with Puppis above them.

Looking south: The two bright stars of Centaurus are high in the southern sky. Achernar may be caught just a few degrees above the southern horizon.

The stars of northern Centaurus are near the zenith.

Key Stars
Canopus: SW, altitude 18°.
Arcturus: N, altitude 34°.
Rigil Kentaurus: S, altitude 61°.
Antares: E, altitude 50°.
Spica: N, altitude 65°.

Magnitude Test Charts

Northern hemisphere: Find Cor Caroli, the star marked in Canes Venatici (high in the southwest on the southern-facing chart), and use the field shown on page 144 (Cor Caroli marks the field centre).

Southern hemisphere: Find Spica (high in the north on the northern-facing chart), and use the field shown on page 144 (Spica marks the field centre).

For Northern Observers:
Boötes (The Herdsman) and Corona Borealis (The Northern Crown)

Boötes, the Herdsman, contains the third-brightest star in the sky, Arcturus, a 'yellow giant' star about a hundred times as luminous as the Sun. It lies some 35 light-years away. These two constellations lie some distance away from the direction of the Milky Way, which marks the plane of the Galaxy. Since clusters and nebulae tend to lie near the Galactic plane and the Milky Way, there are not many examples of these objects within the boundaries of Boötes and Corona Borealis, although there are other interesting things to be seen.

You will find Corona Borealis, the Northern Crown, easy to recognize, because its stars, although not particularly bright, form a distinctive semicircle.

This chart also includes the 'head' of the Serpent (Serpens Caput), which is divided from its body (Serpens Cauda) by the figure of Ophiuchus (see notes for July).

Double stars
(Boötes)
Epsilon (ε) 3.0, 6.3; 2.8″; 340. This celebrated pair will probably prove too difficult for a 60-mm telescope, unless it is exceptionally good, you are a first-class observer, and conditions are excellent! With a

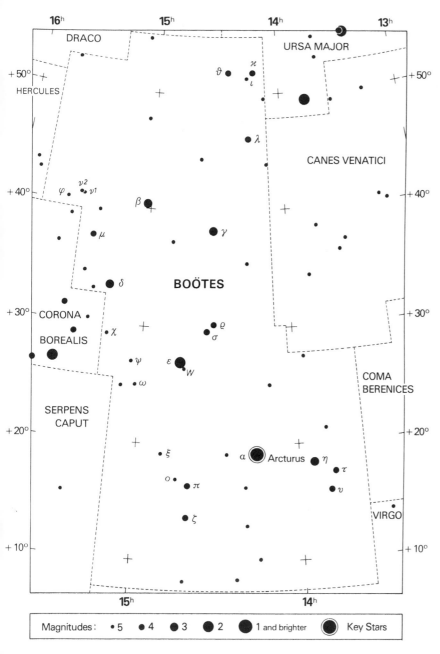

large aperture, the stars are seen to have tints of yellow and green, which have led to its title of *Pulcherrima* (the most beautiful of all). The greenish tint is an effect of contrast, for no such thing as a green star really exists. (T)

Iota (ι) 4.9, 7.5; 38″; 33. White: wide and easy. The companion may be detectable with powerful binoculars. (B/T)

Kappa (κ) 5.1, 7.2; 13″; 237. A pretty pair, white and bluish. (T)

Xi (ξ) 4.8, 6.9; 7.0″; 344. A pretty pair, yellowish and bluish. (T)

Pi (π) 4.9, 6.0; 5.8″; 110. A fine white pair. (T)

(Corona Borealis)

Zeta (ζ) 4.0, 4.9; 6.3″; 306. One of the finest northern pairs. The white primary contrasts beautifully with the apparent blue-green tint of the companion. (T)

Sigma (σ) 5.7, 6.7; 6.3″; 231. A very attractive pair. The primary is yellowish. (T)

(Serpens Caput)

Delta (δ) 4.2, 5.2; 3.9″; 179. A superb pair of white stars. (T)

Variable star

(Corona Borealis)

R A most unusual variable star. Most of its life is spent at a magnitude of about 5.8 – just visible with the naked eye in a very good sky, and readily seen with the smallest optical aid. However, after months or even years of such constancy, it can drop in brightness in the space of a week or two to the 12th or 14th magnitude, as little as a

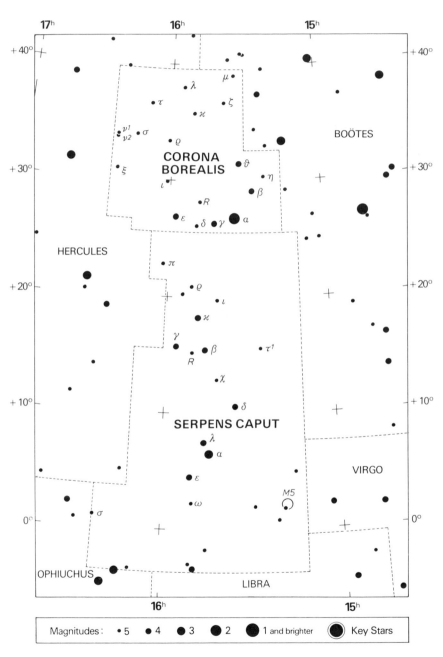

Magnitudes : • 5 • 4 ● 3 ● 2 ⬤ 1 and brighter ◉ Key Stars

thousandth of its normal brightness! Then, after some months – or possibly more than a year – it returns quite quickly to its normal brightness. So it is worth keeping an eye on the little area within the Crown which contains this curious star. (B/T)

For Southern Observers: Scorpius (The Scorpion)

Although Ursa Major, Orion and Crux are the most familiar of all the constellations, it would be easy to make out a case for Scorpius as being the finest of all. Its 'head' contains the brilliant star Antares (literally 'the rival of Mars'), and the magnificent curling trail of bright stars which forms the body and sting ends up in a superb region of the Milky Way. It is so superb because it marks the direction of the centre of our Galaxy. Let a pair of binoculars sweep slowly across the star-clouds of southern Scorpius and you will enjoy a never-to-be-forgotten experience.

However, you need to have Scorpius high in the sky (as it should be, now), and the sky must be dark – it is probable that you will not see these wonders in anything like their true form from an urban garden. But if the conditions are right, you will really see something!

Double stars
Beta (β) 2.6, 4.9; 14″; 23. A splendid white pair. (T)

Nu (ν) 4.0, 6.5; 41″; 336. The primary is yellowish. Divisible with firmly-held and good-quality binoculars. (B/T)

Xi (ξ) 4.8, 7.2; 7.4″; 62. White and bluish. (T)

Sigma (σ) 2.9, 7.8; 20″; 272. The primary is yellow. (T)

Star clusters
M4 A very large, but not particularly bright, globular cluster. Appears hazy in binoculars. (B/T)

M6 A naked-eye cluster. A beautiful curved arrangement of stars. (NE)

M7 Also visible with the naked eye. A bright cluster in a glorious region. (NE)

M80 A small, but very bright, globular cluster. Appears as a hazy blur with small instruments. (B)

NGC 6231 Visible with the naked eye, and attractive even with a small telescope. It lies in a splendid cloud of stars. (NE)

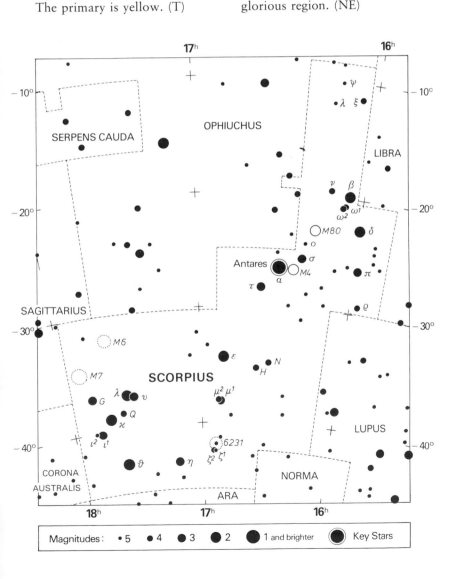

Magnitudes: • 5 • 4 ● 3 ● 2 ● 1 and brighter ⊙ Key Stars

The June Night Sky

(Jun 1, 11 pm; Jun 15, 10 pm; Jun 30, 9 pm)

Northern hemisphere looking north – June

Northern hemisphere looking south – June

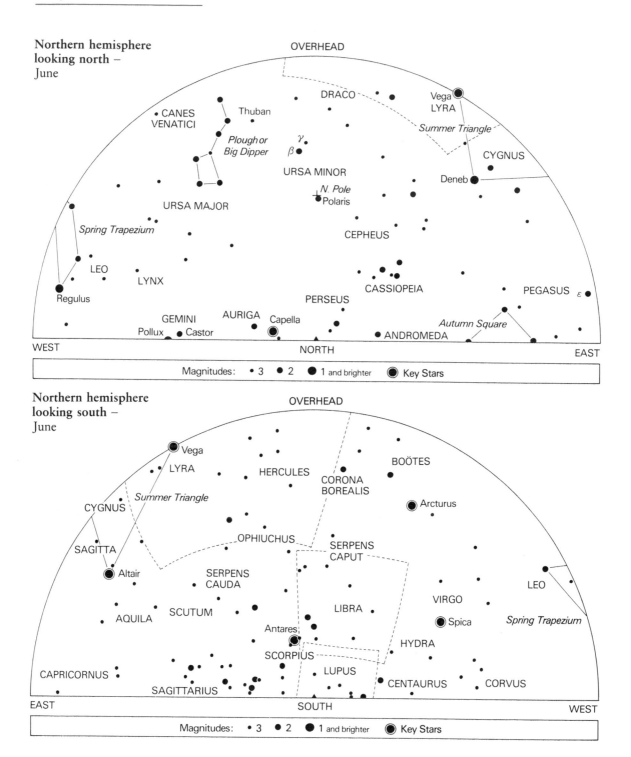

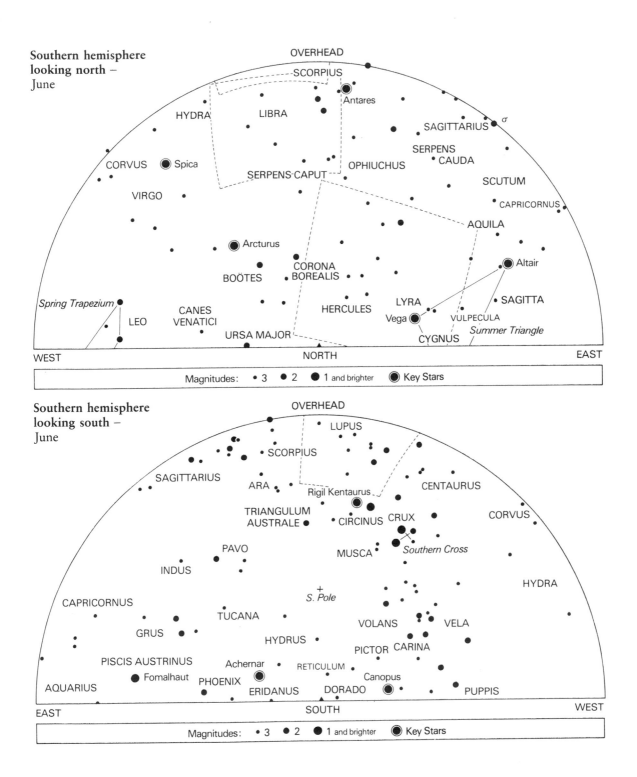

Southern hemisphere looking north – June

OVERHEAD

SCORPIUS

Antares

HYDRA LIBRA

SAGITTARIUS σ

SERPENS CAUDA

CORVUS Spica

SCUTUM

SERPENS·CAPUT OPHIUCHUS

CAPRICORNUS

VIRGO

AQUILA

Arcturus

Altair

CORONA BOREALIS

BOÖTES

SAGITTA

LYRA

HERCULES

Vega

VULPECULA

CANES VENATICI

LEO

Spring Trapezium

Summer Triangle

URSA MAJOR

CYGNUS

WEST NORTH EAST

Magnitudes: • 3 ● 2 ● 1 and brighter ◉ Key Stars

Southern hemisphere looking south – June

OVERHEAD

LUPUS

SCORPIUS

SAGITTARIUS

ARA

CENTAURUS

Rigil Kentaurus

CIRCINUS CRUX CORVUS

TRIANGULUM AUSTRALE

MUSCA

Southern Cross

PAVO

INDUS

HYDRA

S. Pole

CAPRICORNUS

VOLANS VELA

TUCANA

GRUS

HYDRUS

PICTOR CARINA

PISCIS AUSTRINUS

Achernar RETICULUM

Fomalhaut

Canopus

AQUARIUS

PHOENIX

ERIDANUS DORADO

PUPPIS

EAST SOUTH WEST

Magnitudes: • 3 ● 2 ● 1 and brighter ◉ Key Stars

Northern Hemisphere
(Latitude 45°)

Looking east: The famous Summer Triangle, consisting of the stars Vega, in Lyra, Deneb, in Cygnus, and Altair, in Aquila, with the Milky Way passing through it, is now high in the sky. Below them, you may catch the single star Epsilon (ε) Pegasi (magnitude 2.4) not far above the horizon.

Looking south: The bright star Antares, in Scorpius, is now prominent in the lower sky, and nearby Libra is well above the horizon, with Serpens above them and Corona Borealis higher still.

Looking west: Regulus, in Leo, shines conspicuously above the horizon.

Looking north: The stars Beta (β) and Gamma (γ) in Ursa Minor are now vertically above the Pole Star, together with some of the brightest stars of Draco. Perseus is below the Pole, and very near the horizon.

The interesting constellation of Hercules is approaching the zenith.

Key Stars
Arcturus: SW, altitude 59°.
Vega: E, altitude 55°.
Altair: E, altitude 23°.
Antares: S, altitude 17°.
Spica: SW, altitude 27°.

Southern Hemisphere
(Latitude 35°)

Looking east: Sagittarius is now very high in the east, but the sky nearer the horizon is rather empty, though Fomalhaut, in Piscis Austrinus, may be caught just above the horizon.

Looking north: Serpens Caput is on the meridian in mid-sky, with Corona Borealis below it.

Looking west: There are few bright stars in this direction.

Looking south: Look below the Pole to see the bright stars Achernar and Canopus at similar altitudes above the horizon.

The magnificent constellation of Scorpius (leader, Antares) is almost in the zenith.

Key Stars
Arcturus: NNW, altitude 33°.
Rigil Kentaurus: S, altitude 62°.
Altair: ENE, altitude 14°.
Antares: E, altitude 74°.
Spica: NW, altitude 52°.

Magnitude Test Charts

Northern hemisphere: Find Thuban or Alpha (α) Draconis (high in the northwest on the northern-facing chart), and use the field shown on page 162 (Thuban marks the field centre).

Southern hemisphere: Find Sigma (σ) Sagittarii (high in the east on the northern-facing chart), and use the field shown on page 162 (Sigma marks the field centre).

For Northern Observers: Hercules

Hercules is not a conspicuous constellation. Its brightest star, Alpha (α), is only of magnitude 3.3, and it lies near the southern border of the group, near Alpha Ophiuchi. The most helpful guide to Hercules is its trapezium of four stars Epsilon (ε), Zeta (ζ), Eta (η) and Pi (π), often referred to as the *Keystone*. To find the Keystone, cast a line from Vega to Arcturus (which is not shown in the close-up view but may be found on the all-sky map) and pause at a third of the distance along it. Here are some fine objects.

Double stars

Alpha (α) 3.5, 6.1; 4.4″; 110. A superb pair. The primary (a star whose magnitude varies very slightly) is deep gold, making the companion appear green by contrast. You will need to use a high magnification. (T)

Kappa (κ) 5.0, 6.0; 29″; 14. A lovely pair, yellow and deep yellow, and easy with a small telescope. (T)

Rho (ρ) 4.0, 5.1; 3.8″; 317. Close, attractive: both stars white. (T)

95 4.9, 4.9; 6.2″; 259. A magnificent pair, appearing yellowish and white. (T)

100 5.9, 5.9; 14″; 183. Another fine pair, both stars appearing white. (T)

Star clusters

Hercules contains two well-known examples of the globular clusters that are scattered around the Galaxy in a shell or halo.

M13 The brightest globular cluster in the northern sky; in good conditions it can just be detected, as a hazy star, without a telescope. With a low magnification it looks like a hazy patch, but in a very dark sky, and with high magnification, you may be able to glimpse a few stars sparkling in it. Most observers are disappointed with this object in anything less than a 300-mm telescope, but be patient with it, and you may find more rewards than at first seemed likely. It is a long way away – some 22,000 light-years, which is almost as far as the centre of our Galaxy – and it may contain as many as half a million red giant stars, all much more luminous than the Sun! (B)

M92 This is much fainter than M13, and is not so easy to find. The best way may be to use

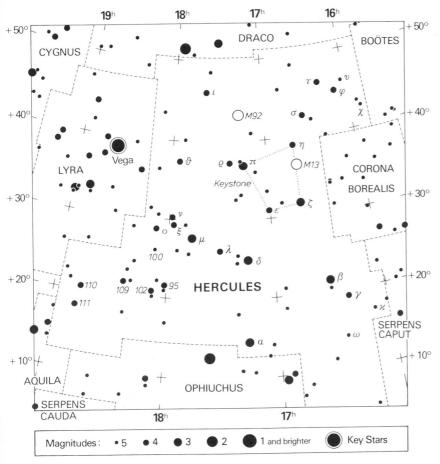

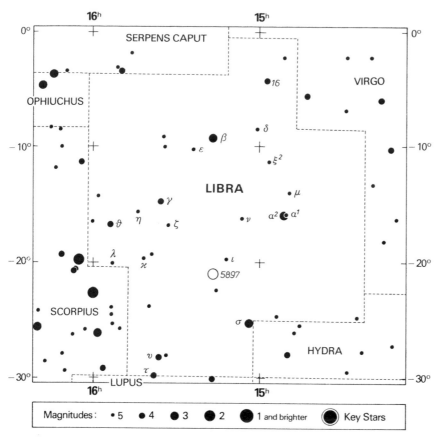

Magnitudes: •5 •4 ●3 ●2 ⬤1 and brighter ◉ Key Stars

seen even with a small telescope. (T)

Mu (μ) 5.4, 6.3; 2.0″; 350. A neat, rather close pair, which should be resolvable with a good 60-mm telescope. (T)

Star cluster

NGC 5897 A globular cluster. The stars are too faint to be resolved with a small telescope, and the object appears as a dim nebulosity. (T)

For Southern Observers:
Lupus (The Wolf)

This constellation is not well known to non-astronomers, but it contains a good sprinkling of fairly bright stars, and lies at the edge of a bright section of the Milky Way.

Double stars

Epsilon (ε) 4.0, 9.0; 26″; 175. The distant companion, though faint, should be readily detectable with a 60-mm telescope. (T)

Kappa (κ) 4.1, 6.0; 27″; 144. A very wide and easy pair; powerful binoculars may be able to resolve it. (B/T)

Mu (μ) 4.5, 7.2; 24″; 130. A very easy double. If you have a telescope of about 120 mm in aperture, look at the brighter star with a very high magnification, as this is in fact a double star itself – see if you can divide it into its two components (5.1, 5.2; 1.0″; 130). (T)

Pi (π) 4.7, 4.8; 1.6″; 65. This is a splendid binary pair, and an

binoculars, which will show it as a small haze, before turning the telescope on to it. Its centre, you will notice, is much more condensed than that of M13. This cluster is believed to be nearly 40,000 light-years away from us, and it is one of the most distant objects that you can see with binoculars or a small telescope within our Galactic system. (T)

For All Observers:
Libra (The Scales)

This constellation is marked by only two moderately bright stars, one of which, Alpha (α), lies almost exactly on the ecliptic. The Sun passes directly in front of this star round about November 7 each year.

Double stars

Alpha (α) 2.8, 5.2. These stars are wide apart even with binoculars. (B)

Iota (ι) 4.7, 9.4; 57.8″; 111. The companion should be readily

excellent test object for a 75-mm telescope. (T)

Star clusters

NGC 5822 A small cluster of faint stars, lying in the Milky Way. There are other clusters within a few degrees of it. (T)
NGC 5986 A globular cluster. The stars are closely packed, and it looks nebulous in a small telescope. (B)

(Jul 1, 11 pm; Jul 15, 10 pm; Jul 31, 9 pm)

Northern Hemisphere
(Latitude 45°)

Looking east: The Autumn Square or Great Square of Pegasus (Alpha (α), Beta (β), and Gamma (γ) Pegasi, and Alpha (α) Andromedae) is now a distinctive sight, but be prepared for it to appear much larger in the sky than you might expect from looking at the chart in this book. To its right, Epsilon (ε) Pegasi sign-posts the tiny constellation of Equuleus, the Little Horse.

Looking south: The splendid constellations of Scorpius and Sagittarius are low in the sky on either side of the meridian.

Looking west: Leo and Virgo are setting, and Arcturus is now high in the western sky.

Looking north: If you have a very low horizon you may just catch Alpha (α) Persei (Mirfak) a little to the right of the north point.

Brilliant Vega (Alpha (α) Lyrae) is very near the zenith.

Key Stars
Arcturus: WSW, altitude 42°.
Vega: SE, altitude 75°.
Altair: SE, altitude 42°.
Antares: SSW, altitude 17°.
Spica: WSW, altitude 11°.

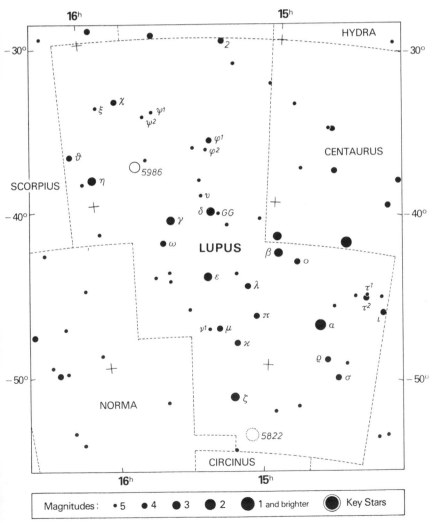

Northern hemisphere looking north – July

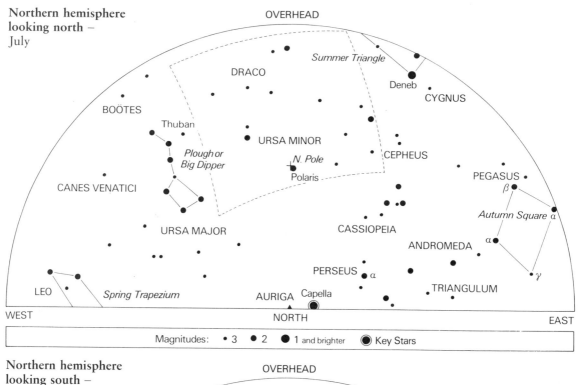

Magnitudes: • 3 ● 2 ⬤ 1 and brighter ◉ Key Stars

Northern hemisphere looking south – July

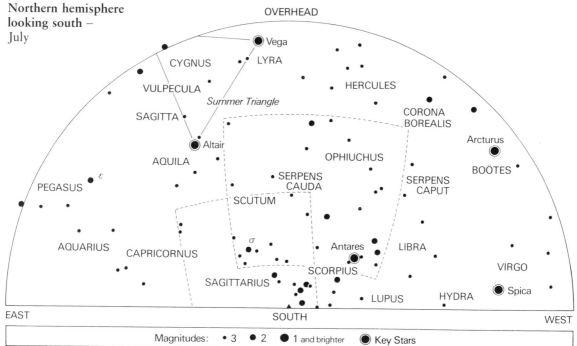

Magnitudes: • 3 ● 2 ⬤ 1 and brighter ◉ Key Stars

Southern hemisphere looking north – July

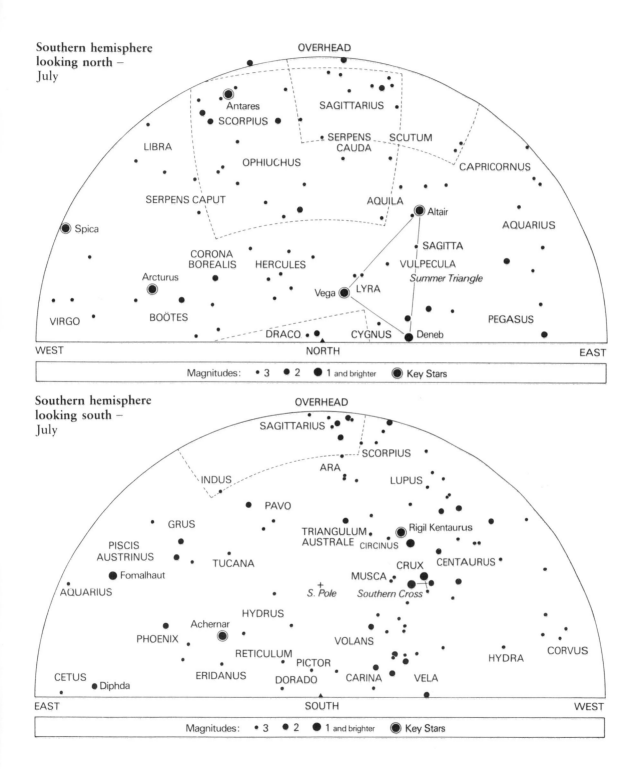

OVERHEAD

SAGITTARIUS

SCORPIUS
Antares

SERPENS CAUDA

SCUTUM

LIBRA

OPHIUCHUS

CAPRICORNUS

SERPENS CAPUT

AQUILA

Altair

AQUARIUS

Spica

SAGITTA

VULPECULA

Summer Triangle

CORONA BOREALIS

HERCULES

Arcturus

Vega LYRA

PEGASUS

VIRGO

BOÖTES

DRACO

CYGNUS

Deneb

WEST

NORTH

EAST

Magnitudes: • 3   ● 2   ● 1 and brighter   ◉ Key Stars

Southern hemisphere looking south – July

OVERHEAD

SAGITTARIUS

SCORPIUS

ARA

INDUS

LUPUS

PAVO

GRUS

TRIANGULUM AUSTRALE

Rigil Kentaurus

CIRCINUS

CENTAURUS

PISCIS AUSTRINUS

TUCANA

CRUX

MUSCA

Fomalhaut

+ S. Pole

Southern Cross

AQUARIUS

HYDRUS

Achernar

PHOENIX

VOLANS

HYDRA

CORVUS

RETICULUM

CETUS

Diphda

PICTOR

ERIDANUS

DORADO

CARINA

VELA

EAST

SOUTH

WEST

Magnitudes: • 3   ● 2   ● 1 and brighter   ◉ Key Stars

Southern Hemisphere
(Latitude 35°)

Looking east: Fomalhaut (Alpha in Piscis Austrinus) dominates the mid-sky. Beta Ceti (Diphda), magnitude 2.0, lies below it.

Looking north: The Milky Way, where it runs from Scorpius to Aquila, crosses the meridian. Note Vega, in Lyra, very low in the north.

Looking west: Corvus, and Spica in Virgo, are now setting.

Looking south: If you have an excellent horizon, try to see Canopus skirting the lower fringes of the sky beneath the celestial pole, with the bright stars of Carina to its right. (Canopus is not included on the hemisphere chart because it is so low in the sky.)

The brilliant star-clouds of Sagittarius are in the zenith, and dominate all else on a dark night.

Key Stars
Arcturus: NW, altitude 19°.
Vega: N, altitude 14°.
Rigil Kentaurus: SW, altitude 52°.
Achernar: SSE, altitude 14°.
Altair: NE, altitude 34°.
Antares: NW, altitude 74°.
Spica: W, altitude 30°.

Magnitude test chart for northern hemisphere – July

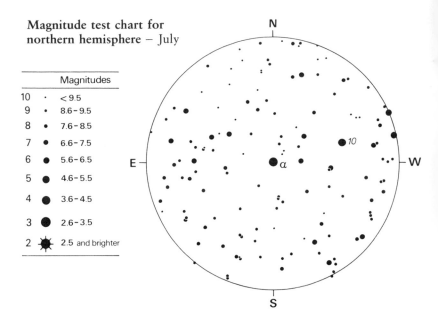

Magnitudes		
10	·	< 9.5
9	·	8.6 - 9.5
8	·	7.6 - 8.5
7	●	6.6 - 7.5
6	●	5.6 - 6.5
5	●	4.6 - 5.5
4	●	3.6 - 4.5
3	●	2.6 - 3.5
2	✳	2.5 and brighter

Magnitude test chart for southern hemisphere – July

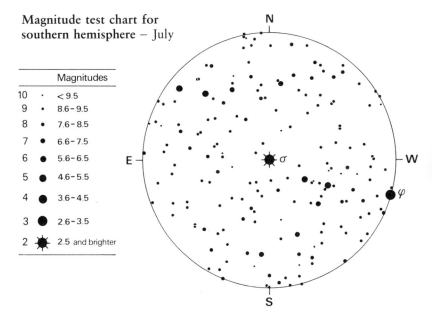

Magnitudes		
10	·	< 9.5
9	·	8.6 - 9.5
8	·	7.6 - 8.5
7	●	6.6 - 7.5
6	●	5.6 - 6.5
5	●	4.6 - 5.5
4	●	3.6 - 4.5
3	●	2.6 - 3.5
2	✳	2.5 and brighter

Magnitude Test Charts

Northern hemisphere: Alpha (α) Draconis (Thuban) marks the field centre.

Southern hemisphere: Sigma (σ) Sagittarii marks the field centre.

For Northern Observers: Draco (The Dragon) and Ursa Minor (The Little Bear)

The constellation of Draco causes all sorts of problems in identification, since it winds for more than a semicircle around the Pole Star, and consequently the stars can appear reversed right for left at different times, according to their position in the sky. Ursa Minor, on the other hand, is readily found, since its brightest star Alpha (α), or Polaris, is the Pole Star, and always appears due north in the sky, at an altitude equal to the observer's latitude on the Earth's surface. Both these constellations lie some distance from the Milky Way, and so they do not contain any particularly dense star fields or bright clusters.

Double stars
(Draco)
Nu (ν) 4.6, 4.6; 62″; 313. A very wide and equal yellowish pair. It looks very pretty in binoculars, which show the two stars well. (B)
Omicron (o) 4.9, 7.9; 35″; 322. A challenge for binoculars, but an easy object with any telescope. (B/T)
Psi (ψ) 4.0, 5.2; 31″; 15. Yellow and purple; a beautiful pair. This may also be divided with good binoculars. (B/T)
17 5.0, 6.0; 3.7″; 116. A rather difficult pair for a small telescope, but an interesting chal-lenge. The stars appear white and bluish. (T)
40 5.4, 6.1; 20″; 234. Yellow, white. An easy pair. (T)
(Ursa Minor)
Alpha (α) 2.0, 9.0; 18″; 218. The Pole Star, or Polaris, is a well-known test object for a small telescope. Some re-markably keen-eyed observers have seen the faint companion with apertures of only 30 mm, but you will do very well to spot it with a 60-mm refractor. It is difficult because the bright star is so brilliant that it causes the fainter one to be lost in the dazzle. (T)

Nebula
(Draco)
NGC 6543 This is a challeng-ing object. It is a planetary nebula – a shell of gas expelled from a dying star, and glowing

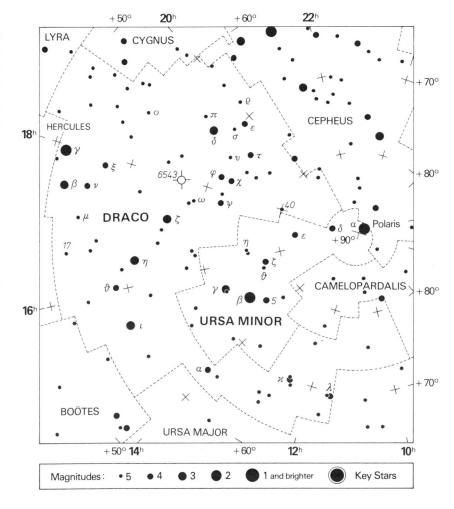

| Magnitudes: | • 5 | ● 4 | ● 3 | ● 2 | ⬤ 1 and brighter | ◎ Key Stars |

brightly. This particular example is bright – about the 6th magnitude – but it is so intense that you will take it for an ordinary star unless you examine it carefully with a powerful eyepiece and see the hazy disc surrounding the nucleus. (T)

For All Observers:
Ophiuchus (The Serpent Bearer) and Serpens Cauda (The Serpent's Body)

Ophiuchus is one of the larger constellations, but its stars do not form a very conspicuous pattern, since they are distributed around its borders. The northern part, where Alpha (α), with a magnitude of 2.1, is found, is not very distinguished, but the southern extremity reaches the magnificent region of the Milky Way near the Galactic centre, between Scorpius and Sagittarius. Serpens Cauda, the body of the Serpent, lies on its eastern border.

Double stars
(Ophiuchus)
Omicron (o) 5.5, 6.0; 11″; 355. Yellow and blue. Fine contrast. (T)
Tau (τ) 5.3, 6.0; 1.9″; 273. A very hard test object for a 60-mm telescope, and you will need the highest magnification available, as well as very steady air. (T)
61 5.5, 5.8; 21″; 94. A splendid white pair. (T)

70 4.3, 6.0; 3.4″; 78. Both stars are deep yellow. (T)
(Serpens Cauda)
Theta (θ) 4.5, 4.5; 22″; 104. One of the sky's finest wide double stars. Well seen with any telescope. (T)
59 5.5, 7.8; 3.9″; 317. A small pair. The primary is yellowish. (T)

Star clusters
(Ophiuchus)
M9 A globular cluster, bright, but the stars are individually too dim and close-packed to be

made out in small instruments. (T)
M10 A bright globular, less easily resolved than M12. (B)
M12 A very loose globular cluster. The stars are so scattered that they can be resolved in a small telescope. Easily found. (B)
M19 A condensed globular cluster of faint stars. (T)
NGC 6633 This is an ordinary open cluster of stars, visible with the naked eye. It is a magnificent low-power telescopic sight. (NE)

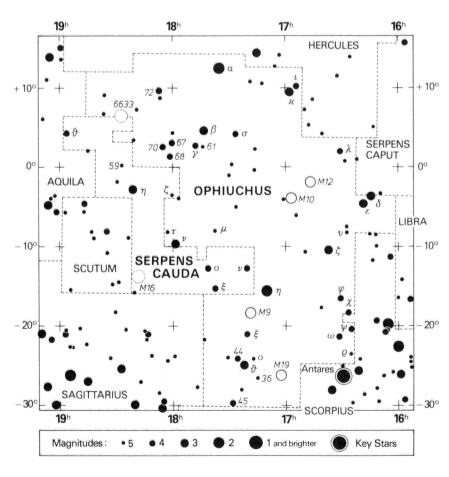

For Southern Observers:
Sagittarius (The Archer)

I have already mentioned Scorpius, the neighbour of Sagittarius and the constellation which lies in the direction of the centre of our Galaxy. However, the star clusters and the Milky Way star-clouds in Sagittarius are even more distinctive than those of the Scorpion. Long-exposure photographs reveal countless millions of stars in seemingly depthless array, and the senses simply cannot appreciate that each one of these specks is probably a mightier star than our own Sun, and separated from its neighbour by the same order of profundity that isolates our own solar system in such frightening solitude! Just sweeping across this region with binoculars or a low-power telescope, when the sky is clear and moonless and the Archer is near the zenith, is enough to awe the most casual skygazer.

Sagittarius contains no outstanding star (the brightest is Epsilon (ε), magnitude 1.8), but the 'bow', which points towards Scorpius, consisting of the stars Gamma (γ), Delta (δ), Epsilon (ε), and Lambda (λ), is always distinctive.

Star clusters

M21 A coarse open cluster of bright stars. (B)

M22 A brilliant globular cluster, obvious to the naked eye. (NE)

M23 A bright open cluster, lying near a 7th-magnitude star. (B)

M28 A very compact globular cluster, appearing rather faint. (T)

M55 A very loose globular cluster, and a 60-mm telescope will detect some individual stars within it. (B)

Nebulae

M8 The *Lagoon Nebula*. A brilliant naked-eye emission nebula. (NE)

M20 The *Trifid Nebula*. A mass of glowing gas and stars, detectable with binoculars. A telescope reveals dark lanes, dividing it up into three main masses. (B)

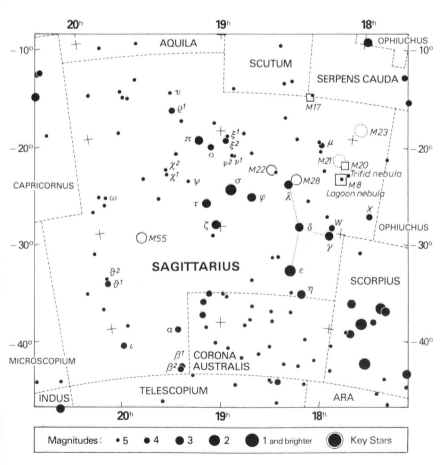

The August Night Sky

(Aug 1, 11 pm; Aug 15, 10 pm; Aug 31, 9 pm)

Northern hemisphere looking north – August

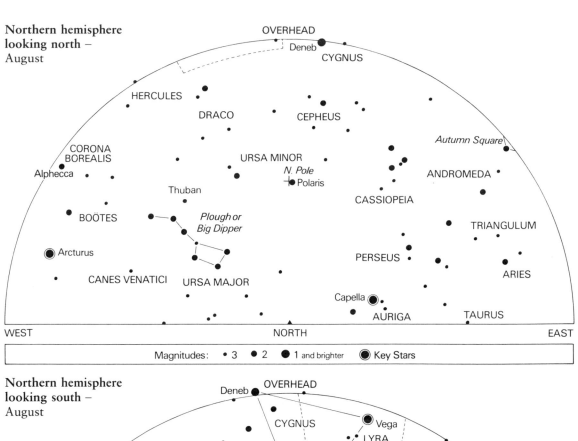

OVERHEAD

Deneb
CYGNUS

HERCULES

DRACO
CEPHEUS

Autumn Square

CORONA BOREALIS
Alphecca

URSA MINOR
N. Pole
Polaris

ANDROMEDA

Thuban
CASSIOPEIA

BOÖTES
Plough or Big Dipper

TRIANGULUM

Arcturus

PERSEUS

CANES VENATICI
URSA MAJOR

ARIES

Capella
AURIGA
TAURUS

WEST · NORTH · EAST

Magnitudes: • 3 ● 2 ● 1 and brighter ◉ Key Stars

Northern hemisphere looking south – August

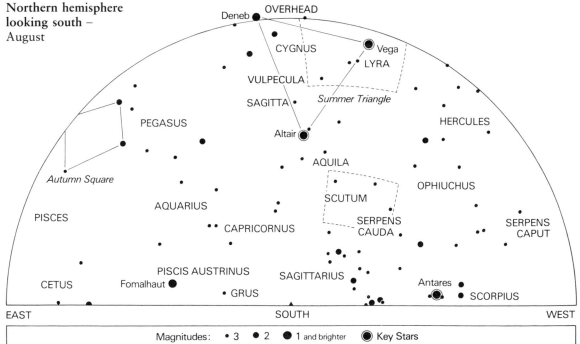

Deneb OVERHEAD

CYGNUS
Vega
LYRA

VULPECULA

SAGITTA
Summer Triangle

PEGASUS
HERCULES

Altair

Autumn Square

AQUILA

OPHIUCHUS

PISCES
AQUARIUS
SCUTUM
SERPENS CAUDA
SERPENS CAPUT

CAPRICORNUS

PISCIS AUSTRINUS

CETUS
Fomalhaut
SAGITTARIUS
Antares
SCORPIUS

GRUS

EAST · SOUTH · WEST

Magnitudes: • 3 ● 2 ● 1 and brighter ◉ Key Stars

Southern hemisphere looking north – August

Southern hemisphere looking south – August

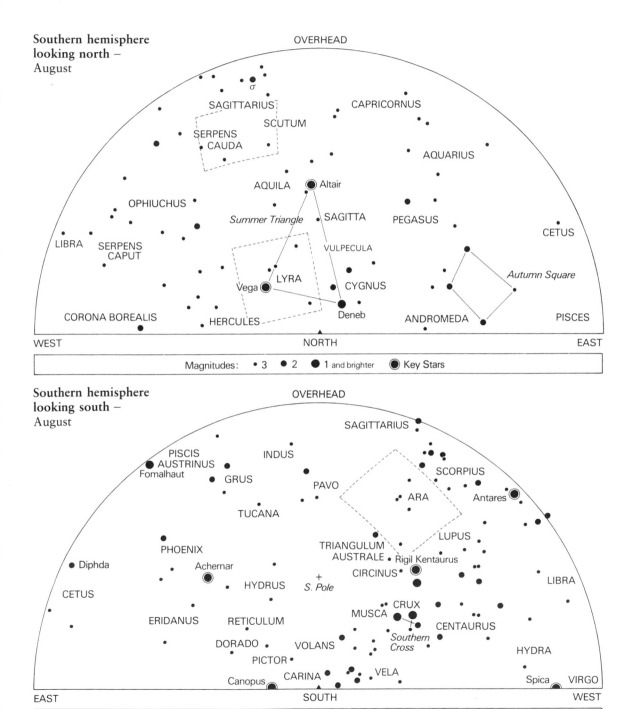

Northern Hemisphere
(Latitude 45°)

Looking east: Andromeda is in mid-sky, while Aries and Triangulum lie below.

Looking south: Aquila, with its bright star Altair, is distinctive, with a rich region of the Milky Way running past it.

Looking west: Arcturus, in Boötes, is in mid-sky, with Alphecca (Alpha Coronae Borealis) above it.

Looking north: Notice how Ursa Major and Perseus are lower in altitude than the celestial pole, and on opposite sides of the north point on the horizon.

Vega and Deneb, in Lyra and Cygnus respectively, are on opposite sides of the zenith.

Key Stars
Arcturus: W, altitude 21°.
Vega: W, altitude 78°.
Altair: S, altitude 53°.

Southern Hemisphere
(Latitude 35°)

Looking east: Diphda (Beta Ceti) is now fairly high in the sky, with Fomalhaut, in Piscis Austrinus, above it.

Looking north: Altair, in Aquila, is on the meridian, with Vega, in Lyra, and Deneb, in Cygnus, on either side and very near the horizon.

Looking west: Spica, in Virgo, has practically set. High above, Antares, in Scorpius, is beginning to descend towards the western horizon.

Looking south: Canopus, in Carina, and the stars of Vela lie to left and right of the south point, very low down. Higher in the sky, Achernar, in Eridanus, and the twin bright stars of Centaurus lie on either side of the celestial pole, acting as a useful guide.

The region of sky between Sagittarius and Piscis Austrinus now lies in the zenith.

Key Stars
Vega: N, altitude 14°.
Rigil Kentaurus: SW, altitude 38°.
Achernar: SE, altitude 28°.
Altair: N, altitude 45°.
Antares: W, altitude 50°.

Magnitude Test Charts

Northern hemisphere: Find Thuban or Alpha (α) Draconis (in the northwest on the northern-facing chart), and use the field shown on page 162 (Thuban marks the field centre).

Southern hemisphere: Find Sigma (σ) Sagittarii (near the zenith on the northern-facing chart), and use the field shown on page 162 (Sigma marks the field centre).

For Northern Observers: Lyra (The Lyre)

Lyra is a very small constellation, but its brilliant star Vega means that it is always noticeable. In fact, Vega is probably the most distinctive individual star in the northern half of the sky.

Double stars

Beta (β) See below.

Epsilon (ε) This is the famous *Double double*. Binoculars, or a keen unaided eye, reveal two 4th-magnitude stars $3\frac{1}{2}'$ arc apart. With a moderate telescope and magnification, each star turns into a close double, the brighter one 4.9, 5.2; 2.3"; 100, and the fainter one 4.6, 6.3; 2.9"; 355. (B/T)

Zeta (ζ) 4.2, 5.5; 44"; 150. Well seen with binoculars. (B)

Variable star

Beta (β) The magnitude of this star appears to change because it forms a close binary, the individual stars orbiting each other once every 12 days 22 hours. They cannot be seen individually, but when one passes in front of the other the total light seems to drop, from magnitude 3.3 to 4.2. For two or three nights in every thirteen, therefore, Beta appears fainter than normal. Compare its brightness with Gamma (γ) (magnitude 3.3) and Kappa (κ) (magnitude 4.3). (NE/B)

Star cluster

M56 A globular cluster, rather small and dim. The individual stars cannot be seen without using a moderate instrument and high magnification. (T)

Nebula

M57 The famous *Ring Nebula*, a planetary nebula in the form of a tiny 'smoke-ring'. It is detectable with a very small telescope, but a moderate magnification will be needed to show it as anything more than a small disc. (T)

For All Observers:
Scutum (The Shield)

This little constellation lies in the Milky Way, and there are many fine objects and beautiful star fields here. Its stars seem to merge with those of Aquila (see notes for September). On a dark night, Scutum's Milky Way fields are utterly magnificent.

Star clusters

M11 The *Wild Duck* cluster. A most glorious concentration of stars, curiously wedge-shaped – hence the allusion to a flight of ducks – and lying at the heart of a wonderful region of the Milky Way, where stars seem to have been flung in careless heaps upon the sky. The star-cloud containing M11 can easily be seen as a bright nebulosity with the naked eye. For observers located near our northern standard latitude, Scutum offers the finest Milky Way sights – the even more brilliant constellations of Scorpius and Sagittarius are too near the horizon, and hence dimmed by haze, to be well seen. (B/T)

M26 An attractive though small cluster. A few of the stars are much brighter than the rest, and seem to stand out against them. (B/T)

NGC 6682 Two gatherings of dim stars amidst many other Milky Way congregations. (T)

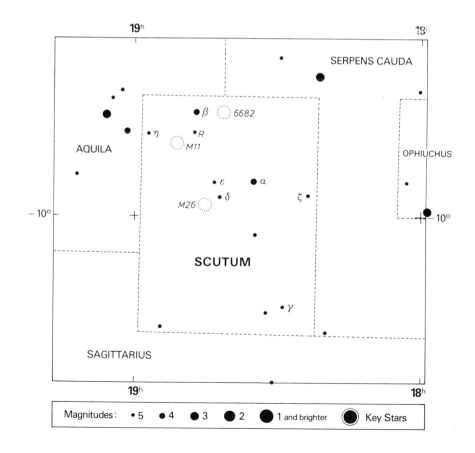

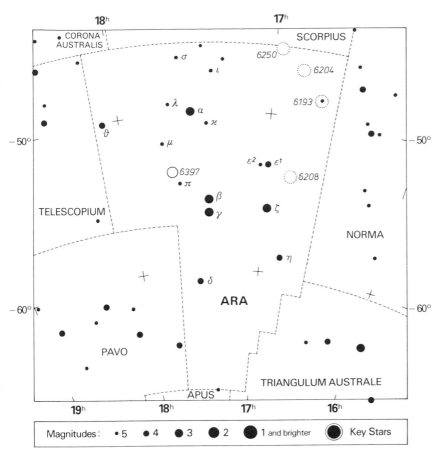

Magnitudes: • 5 ● 4 ● 3 ● 2 ● 1 and brighter ⊙ Key Stars

(Sep 1, 11 pm; Sep 15, 10 pm; Sep 30, 9 pm)

Northern Hemisphere (Latitude 45°)

Looking east: You will see the Pleiades star cluster in Taurus, the sign of approaching winter, well above the horizon, and Aldebaran, the eye of the Bull, is now clearing the horizon. Alpha (α) Ceti (Menkar) may also be made out.

Looking south: Low down, Fomalhaut, in Piscis Austrinus, is approaching the meridian. Aquarius is due south. It contains no very bright star, but may be located by the group of five 3rd- and 4th-magnitude stars two thirds of the way from Fomalhaut to Epsilon (ε) Pegasi.

Looking west: Corona Borealis and Ophiuchus are setting.

Looking north: Note Cepheus, directly above the Pole Star.

Key Stars
Vega: W, altitude 57°.
Capella: NE, altitude 17°.
Altair: SW, altitude 47°.

For Southern Observers: Ara (The Altar)

The Altar is not a well-known constellation, but it contains some fine Milky Way objects, and is well worth sweeping with binoculars, for many attractive fields will be found.

Star clusters
NGC 6193 An open cluster, marked by a 6th-magnitude star, which does not really belong to it, but happens to lie in the same direction. (B)

NGC 6208 This open cluster contains stars of a large range of brightness. (T)

NGC 6250 Another open cluster. The brighter stars can be made out individually in binoculars. (B)

NGC 6397 A globular cluster, visible in a good finder, but not spectacular in a small telescope. (B)

Southern Hemisphere (Latitude 35°)

Looking east: The rather faint group of Eridanus is rising, and contains no particularly bright stars. You may catch Alpha (α) Ceti (Menkar) low down, a little north of the east point.

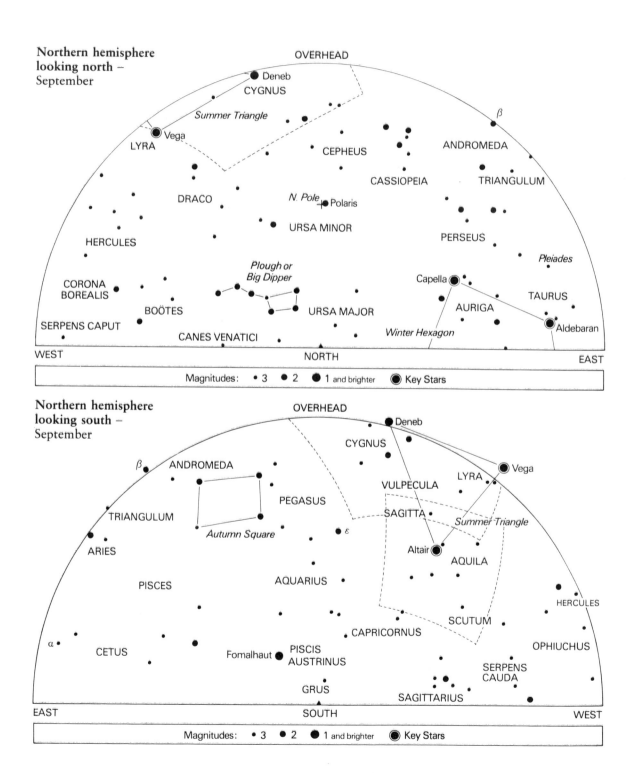

Northern hemisphere looking north – September

OVERHEAD

Deneb

CYGNUS

Summer Triangle

β

Vega

LYRA

CEPHEUS

ANDROMEDA

CASSIOPEIA

TRIANGULUM

DRACO

N. Pole Polaris

URSA MINOR

PERSEUS

HERCULES

Pleiades

Plough or Big Dipper

CORONA BOREALIS

Capella

TAURUS

BOÖTES

URSA MAJOR

AURIGA

SERPENS CAPUT

Aldebaran

CANES VENATICI

Winter Hexagon

WEST

NORTH

EAST

Magnitudes: • 3 ● 2 ⬤ 1 and brighter ◉ Key Stars

Northern hemisphere looking south – September

OVERHEAD

Deneb

CYGNUS

β

ANDROMEDA

Vega

LYRA

VULPECULA

PEGASUS

SAGITTA

Summer Triangle

TRIANGULUM

Autumn Square

ε

ARIES

Altair

AQUILA

PISCES

AQUARIUS

α

SCUTUM

HERCULES

CETUS

CAPRICORNUS

Fomalhaut

PISCIS AUSTRINUS

OPHIUCHUS

SERPENS CAUDA

GRUS

SAGITTARIUS

EAST

SOUTH

WEST

Magnitudes: • 3 ● 2 ⬤ 1 and brighter ◉ Key Stars

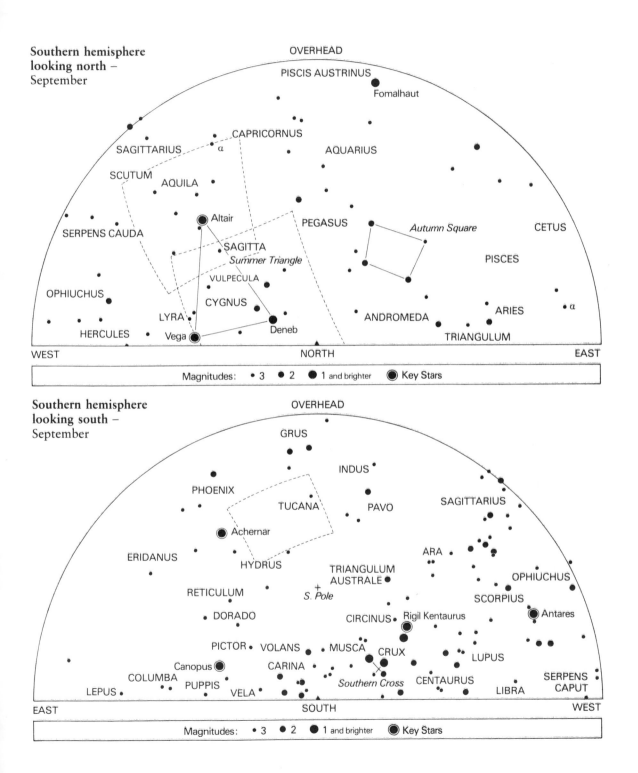

Southern hemisphere looking north – September

OVERHEAD

PISCIS AUSTRINUS

Fomalhaut

CAPRICORNUS

α

SAGITTARIUS

AQUARIUS

SCUTUM

AQUILA

SERPENS CAUDA

Altair

PEGASUS

Autumn Square

CETUS

SAGITTA

Summer Triangle

PISCES

VULPECULA

OPHIUCHUS

CYGNUS

ARIES

α

LYRA

ANDROMEDA

HERCULES

Vega

Deneb

TRIANGULUM

WEST

NORTH

EAST

Magnitudes: • 3 ● 2 ● 1 and brighter ◉ Key Stars

Southern hemisphere looking south – September

OVERHEAD

GRUS

INDUS

PHOENIX

TUCANA

PAVO

SAGITTARIUS

ARA

OPHIUCHUS

ERIDANUS

Achernar

HYDRUS

TRIANGULUM AUSTRALE

RETICULUM

+ S. Pole

SCORPIUS

CIRCINUS

Rigil Kentaurus

Antares

DORADO

PICTOR

VOLANS

MUSCA

CRUX

LUPUS

CANOPUS

CARINA

Southern Cross

CENTAURUS

SERPENS CAPUT

COLUMBA

PUPPIS

VELA

LIBRA

LEPUS

EAST

SOUTH

WEST

Magnitudes: • 3 ● 2 ● 1 and brighter ◉ Key Stars

Looking north: Pegasus is now approaching the meridian, with Aquarius above.

Looking west: The middle sky is dominated by Scorpius and its brilliant star Antares, falling head-first towards the horizon.

Looking south: The brilliant region of Carina and Crux is passing beneath the celestial pole.

Fomalhaut, in Piscis Austrinus, shines near the zenith, passing within 5° of the exact overhead point.

Key Stars
Rigil Kentaurus: SSW, altitude 24°.
Achernar: SE, altitude 44°.
Altair: NNW, altitude 39°.
Antares: W, altitude 25°.

Magnitude Test Charts

Northern hemisphere: Find Beta (β) Andromedae (due east on both charts), and use the field shown on page 180 (Beta marks the field centre).

Southern hemisphere: Find Alpha (α) Capricorni (high in the northwest on the northern-facing chart), and use the field shown on page 180 (Alpha marks the field centre).

For Northern Observers: Cygnus (The Swan) and Vulpecula (The Fox)

Cygnus is an easy constellation to identify at this time, since its leader, Deneb (magnitude 1.3), passes exactly through the zenith, its sky declination being 45°, the same as the standard latitude. Having found Deneb, the wings of the Swan and the long neck leading to the head (Beta (β)) can readily be made out, particularly since Beta lies between the other two bright stars of the Summer Triangle, Vega, in Lyra, and Altair, in Aquila.

The little constellation of Vulpecula is not conspicuous to the naked eye, but it

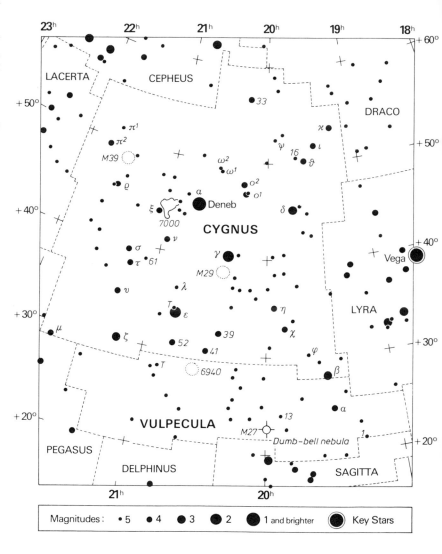

Magnitudes: •5 •4 ●3 ●2 ●1 and brighter ◉ Key Stars

contains the famous *Dumb-bell Nebula*.

Double stars
(Cygnus)

Beta (β) Albireo. 3.0, 5.5; 35″; 55. Wide, easy, and famous. Its colours of rich yellow and blue (the intense-looking blue is a contrast effect) can be detected with a very small telescope: even with binoculars. A splendid double star to show to friends! (B/T)

Omicron (o) A magnitude 3.7 star with a very wide magnitude 4.8 binocular companion. These stars show a lovely contrast of yellow and blue. (B)

Psi (ψ) 5.0, 7.5; 2.9″; 170. A fairly close and delicate pair. (T)

16 5.1, 5.3; 38″; 134. The stars are yellowish. This pair, also, may be divided with carefully-used binoculars. (B/T)

61 5.3, 5.9; 28″; 140. One of the most important stars in the sky. This binary pair was used to make the first accurate measurement of the distance from the Sun to another star, in 1838.* It is only 11 light-years away, and the stars are two of the very few red dwarfs readily visible with binoculars or the naked eye. (NE/B/T)

---

* The distance-measuring technique used the Earth's six-monthly traverse of a diameter of its orbit (300 million km) as a baseline for measuring the tiny shift of 61 Cygni compared with more distant stars in the same field of view. This effect is known as *parallax*. The six-monthly shift of 61 Cygni is equivalent to the thickness of a human hair viewed from a distance of 40 m!

Variable star
(Cygnus)

Chi (χ) This long-period variable takes about 410 days to pass through one complete cycle, and for most of this time it is too faint for effective observation with a small aperture. But near maximum brightness it shines with a distinctive rich red light, usually reaching the 4th or 5th magnitude, but sometimes being considerably brighter than this. So, keep a binocular eye on this space! (NE/B)

Star clusters
(Cygnus)

M29 A rather dim open cluster of telescopic stars. (T)

M39 A brighter cluster of scattered stars. (B)

Nebulae
(Cygnus)

NGC 7000 The famous *North America Nebula*, a glowing patch larger than the apparent diameter of the Moon, but very faint. Choose an extremely clear night, and slowly sweep a pair of binoculars over the region. It will probably be imperceptible using a telescope, since the magnification will be too high to 'condense' its light. (B)

(Vulpecula)

M27 The *Dumb-bell Nebula*. This, probably, is the brightest and finest planetary nebula in the sky. Even with small binoculars, if the sky is dark, you will notice a ghostly disc sail through the field of view as you sweep carefully over the

region. With an aperture of 60 mm, and a low magnification, study it carefully. Unlike most ordinary nebulae, it seems to be as bright at the edge as at the centre. You are looking at a sphere of gas expelled from a dying star in an explosion which must have occurred before recorded history. (B/T)

For All Observers: Aquila (The Eagle) and Sagitta (The Arrow)

These adjoining Milky Way constellations can be well observed from both standard latitudes. Take care to distinguish the little constellation of Sagitta (the Arrow) from its more majestic southern counterpart, Sagittarius (the Archer).

Double stars
(Sagitta)

Zeta (ζ) 5.7, 8.8; 8.5″; 313. A delicate, attractive pair, with colours of white and blue. You may have to spend some time identifying the faint companion. (T)

Theta (θ) 6.0, 8.3; 11″; 327. The primary has a yellowish hue. There is a nearby 7th-magnitude star as well. (T)

(Aquila)

23 5.5, 9.5; 3.4″; 8. A very difficult double for a 60-mm telescope, but worth the attempt, using a high magnification. If you succeed in detecting the companion, both you and your telescope are worth

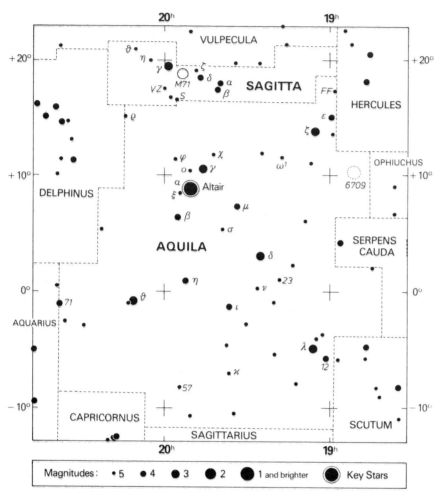

Magnitudes: • 5 • 4 ● 3 ● 2 ● 1 and brighter ⊙ Key Stars

Star cluster
(Sagitta)

M71 An interesting object, though not spectacular in a small instrument, appearing as a hazy, nebulous patch of light with a brighter centre. Astronomers are not sure if this is an unusually condensed normal cluster, or a rather sparsely-populated globular cluster. It can be spotted in binoculars, though it looks extremely dim and is hard to identify. Sagitta is attractive in binoculars, since the whole of the bright 'arrowhead' can be seen almost in the same view. (T)

For Southern Observers:
Tucana (The Toucan)

This constellation is inconspicuous to the naked eye, for its brightest star, Alpha (α), is only of magnitude 2.8. Nevertheless, it has several interesting objects within its boundary.

Double stars

Beta (β) 4.4, 4.5; 27″; 170. Both stars white: a very fine pair, easy with almost any magnification. Good binoculars may elongate or even divide it. (B/T)

Delta (δ) 4.5, 8.1; 7.0″; 282. The primary star is white. A useful test for a small telescope. It might be interesting to experiment and discover the *lowest* magnification which will show the companion. (T)

Kappa (κ) 5.1, 7.3; 5.4″; 346. Primary yellowish. (T)

high praise. If possible, try to view it through a larger telescope first, to have an idea of the relative faintness of the second star. (T)

57 5.2, 6.2; 36″; 171. A wide and easy pair, probably just divisible with binoculars, and obvious with any telescope. (B/T)

Variable star
(Aquila)

Eta (η) This is a pulsating Cepheid-type star (see p. 111), varying from magnitude 4.1 to 5.4 in a period of 7 days 5 hours. Compare its brightness with Iota (ι) (magnitude 4.4); it will usually be the fainter of the two. (NE)

Star clusters

NGC 104 This globular cluster is visible with the naked eye as a blurred 4th-magnitude star. It is often referred to as 47 Tucanae, and is the second most prominent globular in the sky. Only Omega Centauri (see page 147) is larger and brighter. (B/T)

NGC 362 Much fainter than NGC 104, but still a fine object. (T)

Galaxy

The *Small Magellanic Cloud*, or *Nubecula Minor*, is visible with the naked eye as a small grey haze of light. This is a companion galaxy to our own, about 150,000 light-years away. Although less striking than its larger companion in Dorado (see page 194), it contains many telescopic clusters and fine groupings, although a larger aperture than 60 mm will be required to appreciate them fully. (B/T)

The October Night Sky

(Oct 1, 11 pm; Oct 15, 10 pm; Oct 31, 9 pm)

Northern Hemisphere
(Latitude 45°)

Looking east: Aldebaran, in Taurus, is well up in the sky, and Orion is appearing over the horizon – it is always an exciting time for the amateur astronomer, to catch the first view of the Hunter after a summer's absence!

Looking south: The Autumn Square is high up, on the meridian. Fomalhaut, in Piscis Austrinus, shines low down in the sky.

Looking west: Alpha (α) Ophiuchi is now setting, and the bright summer stars Altair, in Aquila, and Vega, in Lyra, are beginning to drop into the western sky.

Looking north: Ursa Major is directly underneath the Pole Star, and Cassiopeia is an equal distance above.

The famous Andromeda Galaxy is in the zenith.

Key stars
Vega: WNW, altitude 36°.
Capella: NE, altitude 32°.
Altair: WSW, altitude 30°.
Aldebaran: E, altitude 21°.

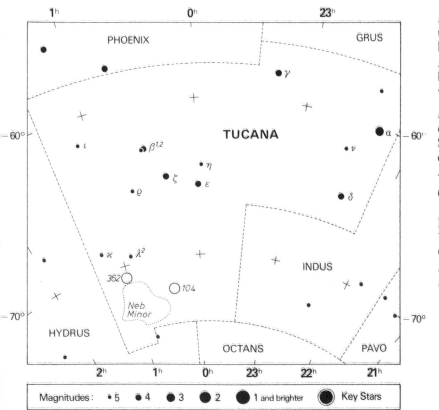

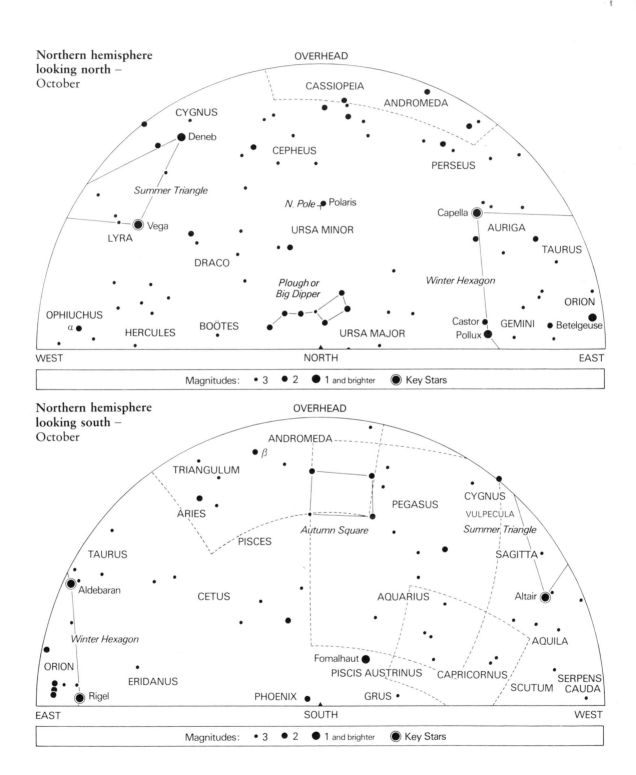

Northern hemisphere looking north – October

OVERHEAD

CASSIOPEIA

ANDROMEDA

CYGNUS

CEPHEUS

Deneb

PERSEUS

Summer Triangle

N. Pole Polaris

Capella

Vega

URSA MINOR

AURIGA

LYRA

TAURUS

DRACO

Plough or Big Dipper

Winter Hexagon

ORION

OPHIUCHUS

α

HERCULES

BOÖTES

URSA MAJOR

Castor

Pollux

GEMINI

Betelgeuse

WEST

NORTH

EAST

Magnitudes: • 3 ● 2 ⬤ 1 and brighter ◉ Key Stars

Northern hemisphere looking south – October

OVERHEAD

ANDROMEDA

β

TRIANGULUM

PEGASUS

CYGNUS

ARIES

VULPECULA

Autumn Square

Summer Triangle

PISCES

SAGITTA

TAURUS

Aldebaran

CETUS

AQUARIUS

Altair

Winter Hexagon

AQUILA

ORION

ERIDANUS

Fomalhaut

PISCIS AUSTRINUS

CAPRICORNUS

SERPENS CAUDA

Rigel

SCUTUM

PHOENIX

GRUS

EAST

SOUTH

WEST

Magnitudes: • 3 ● 2 ⬤ 1 and brighter ◉ Key Stars

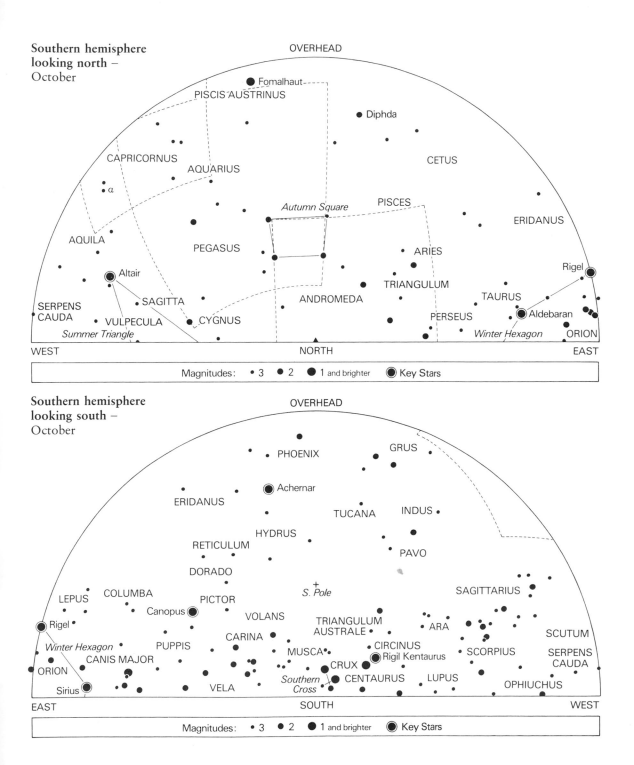

Southern hemisphere looking north – October

OVERHEAD

Fomalhaut
PISCIS AUSTRINUS
Diphda
CETUS
CAPRICORNUS
AQUARIUS
α
ERIDANUS
Autumn Square
PISCES
AQUILA
PEGASUS
ARIES
Altair
TRIANGULUM
Rigel
SAGITTA
TAURUS
SERPENS CAUDA
ANDROMEDA
Aldebaran
VULPECULA
CYGNUS
PERSEUS
Summer Triangle
Winter Hexagon
ORION
WEST
NORTH
EAST

Magnitudes: • 3 • 2 ● 1 and brighter ◉ Key Stars

Southern hemisphere looking south – October

OVERHEAD

PHOENIX
GRUS
Achernar
ERIDANUS
TUCANA
INDUS
HYDRUS
RETICULUM
PAVO
DORADO
S. Pole
SAGITTARIUS
LEPUS
COLUMBA
PICTOR
VOLANS
TRIANGULUM AUSTRALE
ARA
Canopus
SCUTUM
Rigel
CARINA
CIRCINUS
SCORPIUS
Winter Hexagon
PUPPIS
MUSCA
Rigil Kentaurus
SERPENS CAUDA
CANIS MAJOR
CRUX
ORION
Southern Cross
CENTAURUS
LUPUS
OPHIUCHUS
Sirius
VELA
EAST
SOUTH
WEST

Magnitudes: • 3 • 2 ● 1 and brighter ◉ Key Stars

Southern Hemisphere
(Latitude 35°)

Looking east: The sky near the horizon contains many bright stars, with Orion and Canis Major rising.

Looking north: The Autumn Square is low on the meridian, with Diphda (Beta Ceti) higher, and almost due north.

Looking west: Scorpius is setting, with Sagittarius higher and to the right.

Looking south: Carina and Crux make a bright display beneath the celestial pole, and Achernar, in Eridanus, is high above it.

Bright Fomalhaut, in Piscis Austrinus, is slightly west of the zenith.

Key Stars
Canopus: SE, altitude 20°.
Rigil Kentaurus: SSW, altitude 13°.
Achernar: SSE, altitude 58°.
Altair: WNW, altitude 21°.
Aldebaran: E, altitude 24°.

Magnitude Test Charts

Northern hemisphere: Beta (β) Andromedae marks the field centre.

Southern hemisphere: Alpha (α) Capricorni marks the field centre.

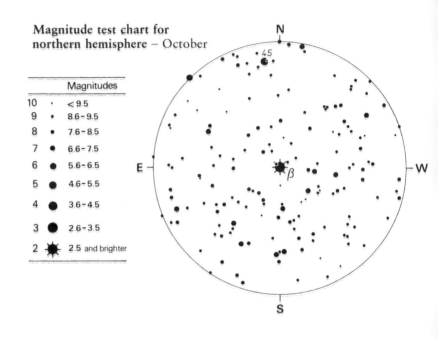

Magnitude test chart for northern hemisphere – October

Magnitudes		
10	·	< 9.5
9	.	8.6 – 9.5
8	•	7.6 – 8.5
7	●	6.6 – 7.5
6	●	5.6 – 6.5
5	●	4.6 – 5.5
4	●	3.6 – 4.5
3	●	2.6 – 3.5
2	✴	2.5 and brighter

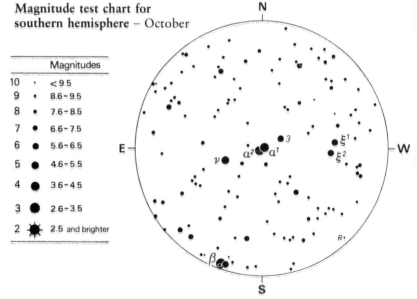

Magnitude test chart for southern hemisphere – October

Magnitudes		
10	·	< 9.5
9	.	8.6 – 9.5
8	•	7.6 – 8.5
7	●	6.6 – 7.5
6	●	5.6 – 6.5
5	●	4.6 – 5.5
4	●	3.6 – 4.5
3	●	2.6 – 3.5
2	✴	2.5 and brighter

For Northern Observers: Andromeda, Pegasus, and Triangulum (The Triangle)

The two large constellations merge, the upper left-hand corner of the so-called Autumn Square being completed by the star Alpha Andromedae (Alpheratz). The Autumn Square is often found difficult to identify, appearing much larger in the sky than it does on the map. It lies in a star-poor region of the sky, and within its extensive area barely a handful of even reasonably conspicuous naked-eye stars are to be found; from most urban regions it will appear entirely blank. The neighbouring small constellation of Triangulum is included because of its interesting galaxy, M33.

Double stars
(Andromeda)
Gamma (γ) 2.3, 5.1; 9.8″; 60. Rich yellow, blue-green. One of the finest pairs in the whole sky. A closer version of Beta (β) Cygni (page 175), but I find Gamma Andromedae to have the finer tints. (T)
Pi (π) 4.1, 8.0; 36″; 173. White primary. This pair is rather too wide for some tastes. (T)
(Pegasus)
I 4.5, 8.6; 36″; 311. The brighter star is a beautiful deep yellow colour. (T)
(Triangulum)
Iota (ι) 5.0, 6.4; 3.9″; 74. A fine pair, appearing yellow and bluish. (T)

Star cluster
(Pegasus)
M15 This bright globular cluster can be seen with binoculars as a hazy disc, and you may be able to spot some individual stars within its boundary, if you use a high magnification with a 60-mm refractor. (B/T)

Galaxies
(Andromeda)
M31 This is the famous *Andromeda Galaxy*, some 2,200,000 light-years away. It is very similar to our own Galaxy. On a reasonably clear night it will be detectable with the naked eye as an elongated haze, and in binoculars it can appear quite large, with a bright nucleus and extensive 'wings' on either side. Look for the two satellite galaxies: one is known as M32, and lies 25′ arc to the south, looking like a small hazy patch. The other, NGC 205, is about 45′ arc northwest, and is harder to identify because, in a small instrument, it appears almost star-like. Many people are disappointed with the Andromeda Galaxy, because it shows little obvious detail. But

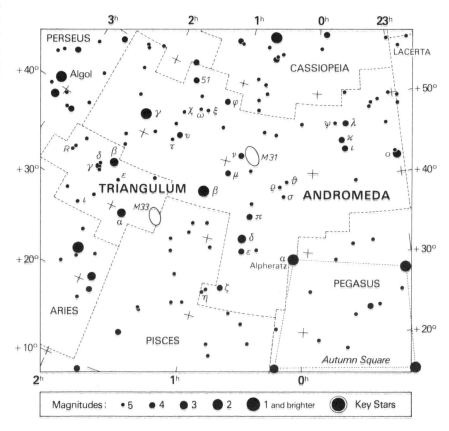

it is amazing to think that this diaphanous ghost contains one hundred thousand million stars, many more luminous than the Sun! (NE/B)

(Triangulum)
M33 Lying only a little further away than M31, this object appears excessively faint, though quite large. Wait for a very dark night, and sweep slowly over the area with binoculars, looking for a 'ghost' passing through the field. It is a spiral galaxy seen almost face-on, and appears beautiful in photographs. (B)

For All Observers:
Aquarius (The Water Carrier)

This constellation is not well known, and it is not easy to identify. The most distinctive part is the little zig-zag of stars south of Pegasus.

Double stars

Zeta (ζ) 4.3, 4.5; 1.8″; 218. This lovely binary is included for the benefit of those observers with apertures of rather more than 60 mm, although if this is the size of your telescope you could attack it with a high magnification and see if the 'star' is elongated in a north-south direction. But it is a severe test indeed, and do not be disappointed if Zeta looks just like an ordinary star. (T)
107 5.3, 6.5; 6.5″; 135. White and bluish; an attractive pair. (T)

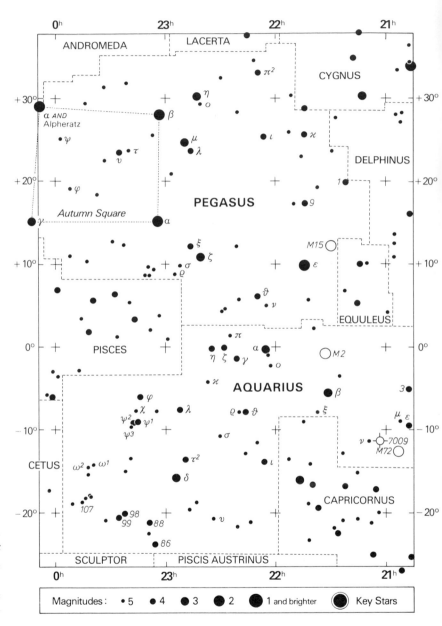

Star cluster

M2 A bright globular cluster, visible with binoculars as a hazy spot. The stars are so densely packed that it still appears hazy when a 60-mm telescope is used. This object is about 55,000 light-years away, making it one of the remotest globular clusters visible with a small instrument. (T)

For Southern Observers:
Capricornus (The Sea Goat)

A large but ill-defined constellation, lying well away from the Milky Way and therefore not possessing any remarkable star fields. It is noteworthy, however, for its naked-eye double star Alpha (α), while Beta (β) has a wide binocular companion of the 6th magnitude.

Double stars
Alpha (α) 3.2, 4.2: 376″ (over 6′ arc); 291. An easy naked-eye double, and very wide and bright in binoculars. (NE)
Delta (δ) 5.5, 9.0; 56″; 180. A very unequal pair, but quite easy with a small telescope. (T)
Pi (π) 5.1, 8.7; 3.4″; 145. Stars yellowish and bluish. (T)

Star cluster
M30 A rather dim, large, globular cluster. It seems to have a much brighter nucleus. Do not expect to see any individual stars with a small aperture, however! This cluster is about 40,000 light-years away. (T)

The November Night Sky

(Nov 1, 11 pm; Nov 15, 10 pm; Nov 30, 9 pm)

Northern Hemisphere (Latitude 45°)

Looking east: The 'twins', Castor and Pollux in Gemini, are well up, and Procyon, in Canis Minor, is also visible.

Looking south: The rather inconspicuous constellation of Cetus is on the meridian, and the distinctive group of three stars in Aries still higher in the sky.

Looking west: The bright star Altair, in Aquila, is setting.

Looking north: The two stars forming the 'body' of Ursa Minor, the Little Bear, are now almost underneath Polaris, the Pole Star.

The fine region of the Milky Way between Cassiopeia and Perseus is now passing near the zenith.

Key Stars
Vega: NW, altitude 18°.
Capella: ENE, altitude 51°.
Rigel: ESE, altitude 17°.
Altair: W, altitude 9°.
Aldebaran: ESE, altitude 42°.

Magnitudes: • 5 • 4 ● 3 ● 2 ● 1 and brighter ◯ Key Stars

Northern hemisphere looking north – November

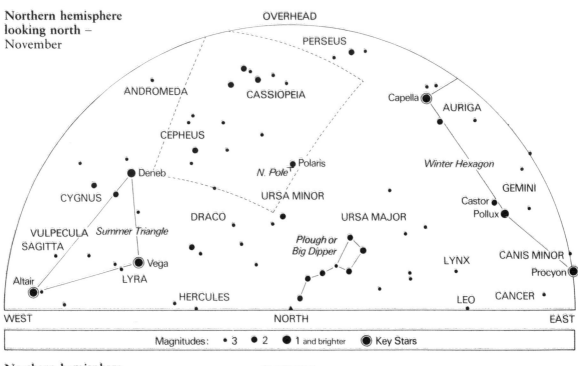

Magnitudes: • 3 ● 2 ⬤ 1 and brighter ◎ Key Stars

Northern hemisphere looking south – November

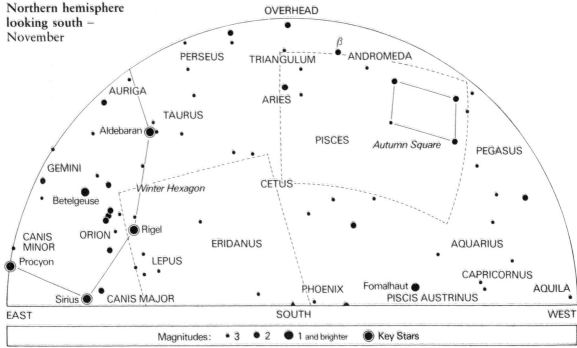

Magnitudes: • 3 ● 2 ⬤ 1 and brighter ◎ Key Stars

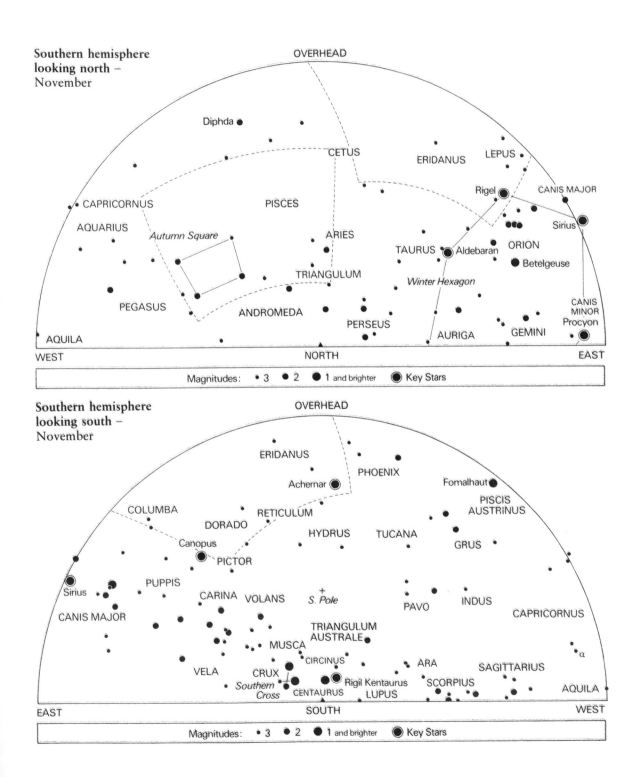

Southern hemisphere looking north – November

OVERHEAD

Diphda

CETUS

ERIDANUS

LEPUS

CAPRICORNUS

AQUARIUS

PISCES

ARIES

Rigel

CANIS MAJOR

Sirius

Autumn Square

TAURUS   Aldebaran

ORION

TRIANGULUM

Betelgeuse

PEGASUS

ANDROMEDA

Winter Hexagon

CANIS MINOR

Procyon

AQUILA

PERSEUS

AURIGA

GEMINI

WEST

NORTH

EAST

Magnitudes: • 3   ● 2   ● 1 and brighter   ◉ Key Stars

Southern hemisphere looking south – November

OVERHEAD

ERIDANUS

PHOENIX

Achernar

Fomalhaut

PISCIS AUSTRINUS

COLUMBA

DORADO

RETICULUM

HYDRUS

TUCANA

GRUS

Canopus

PICTOR

PUPPIS

CARINA

VOLANS

S. Pole

PAVO

INDUS

CAPRICORNUS

Sirius

CANIS MAJOR

TRIANGULUM AUSTRALE

MUSCA

CIRCINUS

ARA

SAGITTARIUS

α

VELA

CRUX

Southern Cross

CENTAURUS

Rigil Kentaurus

LUPUS

SCORPIUS

AQUILA

EAST

SOUTH

WEST

Magnitudes: • 3   ● 2   ● 1 and brighter   ◉ Key Stars

Southern Hemisphere
(Latitude 35°)

Looking east: Sirius, the brilliant star in Canis Major, is well above the horizon, dominating the eastern sky.

Looking north: There are few bright stars in this part of the sky. Note the small constellations of Aries and Triangulum, one above the other, low down.

Looking west: The last bright stars of Sagittarius are setting.

Looking south: The stars of Centaurus and Crux are beneath the celestial pole, with Achernar, in Eridanus, well above it.

The nearest bright star to the zenith is Diphda (Beta Ceti).

Key Stars
Sirius: E, altitude 19°.
Canopus: SE, altitude 37°.
Rigel: ENE, altitude 32°.
Achernar: S, altitude 66°.
Aldebaran: NE, altitude 23°.

Magnitude Test Charts

Northern hemisphere: Find Beta (β) Andromedae (high in the southwest on the south-facing chart), and use the field shown on page 180 (Beta marks the field centre).
Southern hemisphere: Find Alpha (α) Capricorni (near the west on the southern-facing chart), and use the field shown on page 180 (Alpha marks the field centre).

For Northern Observers: Cassiopeia and Cepheus

These two constellations occupy a fine part of the Milky Way, and contain many interesting double stars and star clusters, as well as a very famous variable star. Cassiopeia contains five bright stars in the form of a 'W' or 'M' – it sometimes appears reversed or upside-down, since it swings around the celestial pole in the course of 24 hours, always being at least partly above the horizon.

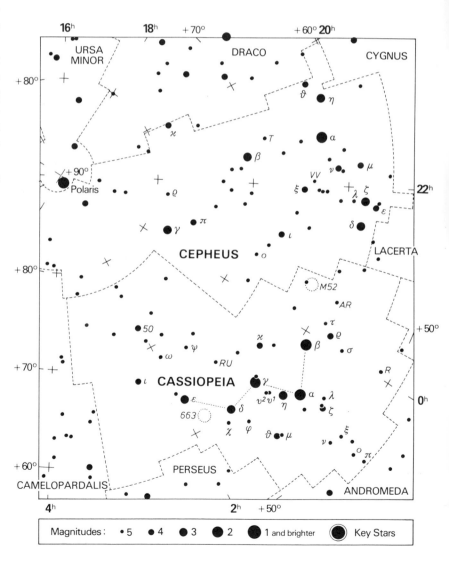

| Magnitudes : | •5 | • 4 | ● 3 | ● 2 | ⬤ 1 and brighter | ◎ Key Stars |

Double stars

(Cassiopeia)

Eta (η) 3.7, 7.4; 11″; 298. A superb pair, yellowish and deep blue. (T)

Iota (ι) AB: 4.2, 7.1; 2.5″; 233. AC: 4.2, 8.1; 7.4″; 113. A magnificent triple star. The bright star (A) is yellow, and the fainter ones look blue. A high magnification is required. (T)

Sigma (σ) 5.4, 7.5; 3.1″; 327. A hard double for a small telescope. (T)

(Cepheus)

Beta (β) 3.3, 8.0; 14″; 250. A large contrast in brightness, but readily seen even with a small telescope. (T)

Delta (δ) 4.0, 5.3; 41″; 192. A very wide and easy pair, with a beautiful contrast of yellow and blue. Divisible with binoculars. (B/T)

Kappa (κ) 4.0, 8.0; 7.4″; 122. A delicate pair. (T)

Xi (ξ) 4.7, 6.5; 8.0″; 280. Yellowish and bluish; a very attractive double. (T)

Variable stars

(Cassiopeia)

Gamma (γ) This is a strange star. Its normal magnitude is 2.3, but in 1937 it brightened to 1.7 – brighter than the Pole Star! It is worth keeping an eye on it, to see if any future change like this occurs. (NE)

(Cepheus)

Delta (δ) The prototype Cepheid variable, changing from magnitude 3.6 to 4.2 and back again in a period of 5 days 9 hours. On some occasions, then, it is as faint as nearby

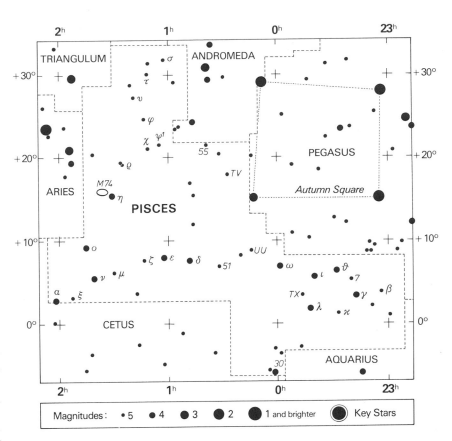

Magnitudes: •5 •4 •3 ●2 ●1 and brighter ⊙ Key Stars

Epsilon (ε), magnitude 4.2, while on others it is as bright as Beta (β), magnitude 3.6. (NE/B)

Mu (μ) A very slowly-changing variable. You will probably not notice much variation, but it is worth identifying for its amazing claret colour. It is known as the Garnet star.

Star clusters

(Cassiopeia)

M52 Visible in binoculars, this open cluster is a fine sight in a small telescope, appearing as a mass of faint stars. (B/T)

NGC 663 A coarse open cluster of bright stars. (B/T)

For All Observers: Pisces (The Fishes)

Pisces is not a distinguished constellation visually, occupying a rather dull area to the south of Andromeda and Pegasus; its brightest star, Alpha (α), is only of magnitude 3.8. However, it is one of the most significant, since the Sun lies in Pisces at the time when it passes from the southern to the northern celestial hemisphere, signifying the beginning of spring in the Earth's northern hemisphere and the beginning of autumn in the southern hemisphere. This moment is known as the *vernal* or *spring equinox*.

Double stars

Alpha (α) 4.2, 5.2; 1.7″; 278. The stars are a curious greenish-white colour. You will not be able to divide them with a 60-mm telescope, but a high magnification may reveal an elongated 'single' star. Try it! (T)

Zeta (ζ) 4.2, 5.3; 24″; 63. Both stars are white. This is a pretty pair, and very easy. (T)

Psi¹ (ψ¹) 4.9, 5.0; 30″; 160. An attractive wide and almost equal pair. They could be divided with good binoculars. (B/T)

55 5.5, 8.2; 6.6″; 193. Fine contrast of yellow and blue. (T)

Galaxy

M74 This is a spiral galaxy, rather dim and showing little detail in a small telescope. You will do well to find it with an aperture of less than about 75 mm, and since the field is readily found, just over one degree east and a little north of Eta (η), it is a good test object. (T)

For Southern Observers: Eridanus (The River Eridanus)

Since Eridanus extends from the celestial equator to a declination of about −60°, parts of it can be well observed from anywhere on the Earth's surface – though not its bright leader, Achernar, which is far south. Gamma (γ), magnitude 3.0 and reddish, is the most conspicuous of its stars as seen from mid-latitudes in the northern hemisphere.

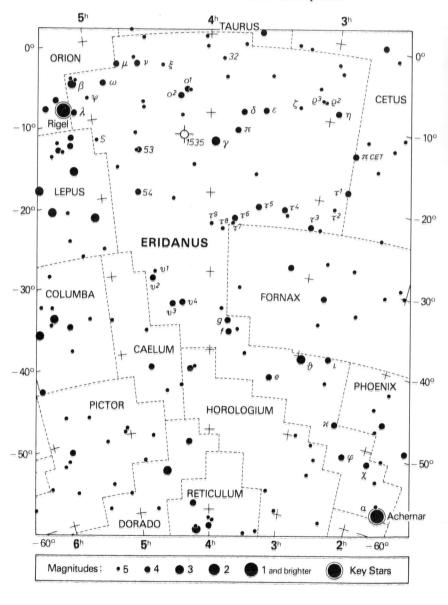

Magnitudes : • 5 • 4 ● 3 ● 2 ● 1 and brighter ⊙ Key Stars

The December Night Sky

(Dec 1, 11 pm; Dec 15, 10 pm; Dec 31, 9 pm)

Double stars

Theta (θ) 3.4, 4.4; 8.2″; 88. A splendid pair of white stars. (T)

Omicron² (o^2) 4.5, 9.7; 83″; 105. Too wide to be an attractive telescopic object, but an interesting test for very powerful binoculars. (B)

f 4.9, 5.4; 7.8″; 211. White. Readily resolved in a small telescope. (T)

32 4.0, 6.0; 7.0″; 347. One of the finest pairs in the sky. The primary is a rich yellow, and the companion appears a bluish-green. (T)

Nebula

NGC 1535 A small planetary nebula, with a brighter centre, showing a tiny disc in a small telescope. (T)

Northern Hemisphere
(Latitude 45°)

Looking east: The head of Leo is rising, and Procyon, in Canis Minor, is well up. Castor and Pollux, in Gemini, are high in the sky.

Looking south: Aldebaran and the Pleiades star cluster, in Taurus, dominate the meridian, with Orion following close behind in the stars' apparent motion across the sky.

Looking west: Pegasus is setting, a sign that autumn is coming to an end.

Looking north: The brightest stars of Draco are beneath the Pole Star.

The fine constellation of Perseus is overhead.

Key Stars
Sirius: SE, altitude 14°.
Capella: E, altitude 71°.
Rigel: SSE, altitude 32°.
Procyon: ESE, altitude 23°.
Aldebaran: SSE, altitude 58°.

Southern Hemisphere
(Latitude 35°)

Looking east: The lower part of the sky is devoid of bright stars, the nearest being those of Vela, towards the celestial pole. Canis Major will be seen high up.

Looking north: Aldebaran, in Taurus, is approaching the northern point, and Capella, in Auriga, may just be caught above the northern horizon.

Looking west: Fomalhaut, in Piscis Austrinus, is shining prominently in mid-sky.

Looking south: The brilliant stretch of the Milky Way from Vela to Centaurus is ranged to the left of the celestial pole.

Key Stars
Sirius: E, altitude 43°.
Canopus: SE, altitude 54°.
Rigel: NE, altitude 54°.
Procyon: ENE, altitude 19°.
Achernar: SSW, altitude 60°.
Aldebaran: N, altitude 36°.

Magnitude Test Charts

Northern hemisphere: Find Capella (east-northeast on the northern-facing chart), and use the field shown on page 127 (Capella marks the field centre).

Southern hemisphere: Find Sirius (east–northeast on the northern-facing chart) and use the field shown on page 127 (Sirius marks the field centre).

**Northern hemisphere
looking north –
December**

**Northern hemisphere
looking south –
December**

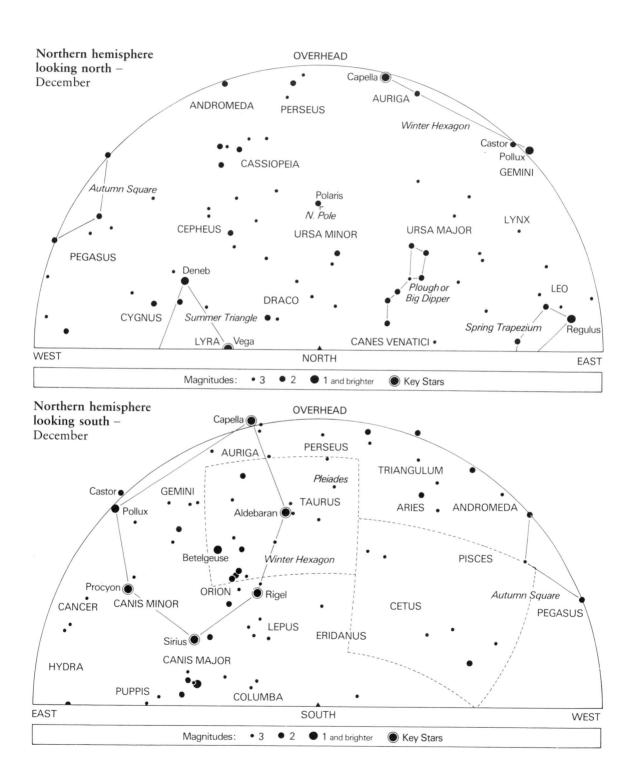

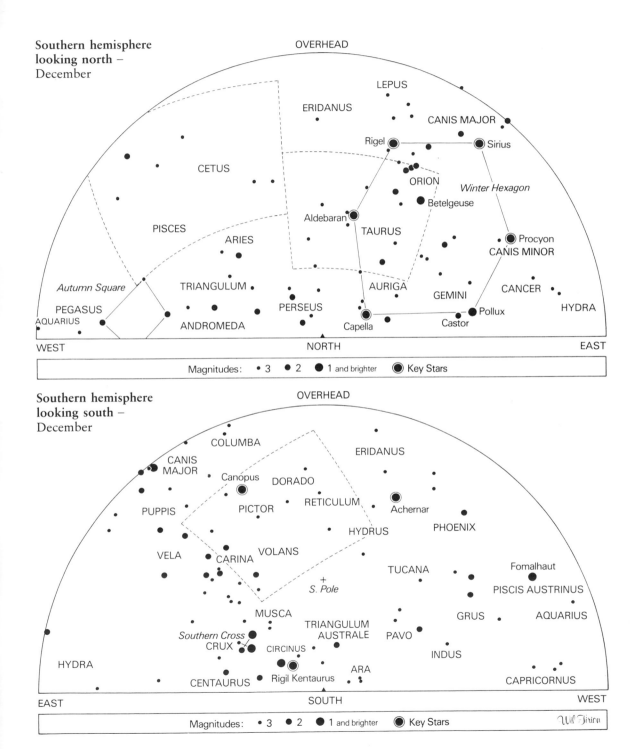

Southern hemisphere looking north – December

OVERHEAD

LEPUS

ERIDANUS

CANIS MAJOR

CETUS

Rigel

Sirius

ORION

Winter Hexagon

Betelgeuse

PISCES

Aldebaran

TAURUS

Procyon

ARIES

CANIS MINOR

CANCER

TRIANGULUM

AURIGA

Autumn Square

GEMINI

HYDRA

PEGASUS

Pollux

AQUARIUS

ANDROMEDA

PERSEUS

Capella

Castor

WEST

NORTH

EAST

Magnitudes: • 3 ● 2 ⬤ 1 and brighter ◎ Key Stars

Southern hemisphere looking south – December

OVERHEAD

COLUMBA

ERIDANUS

CANIS MAJOR

Canopus

DORADO

PICTOR

RETICULUM

Achernar

PUPPIS

HYDRUS

PHOENIX

VELA

CARINA

VOLANS

TUCANA

Fomalhaut

+ S. Pole

PISCIS AUSTRINUS

MUSCA

TRIANGULUM AUSTRALE

GRUS

AQUARIUS

Southern Cross

CRUX

CIRCINUS

PAVO

INDUS

HYDRA

CENTAURUS

Rigil Kentaurus

ARA

CAPRICORNUS

EAST

SOUTH

WEST

Magnitudes: • 3 ● 2 ⬤ 1 and brighter ◎ Key Stars

Wil Tirion

For Northern Observers: Taurus (The Bull)

This constellation is easy to distinguish, with Aldebaran, the Bull's red eye, confronting Orion from the west. It contains two famous naked-eye star clusters, the *Pleiades* and the *Hyades*, the latter containing several of its brighter stars.

Double stars

Alpha (α) 0.8, 11.3; 128″; 33. The companion is a good test object for an aperture of 75 mm. Brilliant Aldebaran tends to dazzle the eye so much that little else can be seen; try moving the telescope so that it is just outside the field of view! (T)

Phi (φ) 5.1, 8.5; 51″; 255. A wide, very easy pair. The brighter star is yellow. It would be interesting to tackle this pair with powerful binoculars, such as 10 × 50s. (B/T)

Chi (χ) 5.7, 7.8; 20″; 25. The primary is white. (T)

Variable star

Lambda (λ) An eclipsing binary star. The magnitude range is from 3.5 to 4.0, and the period is 3 days 23 hours. For comparison stars you could use Epsilon (ε), magnitude 3.5, and Nu (ν), magnitude 3.9. Eclipsing stars repeat their cycles of change very accurately. (NE/B)

Star clusters

M45 The Pleiades, a young, nearby cluster (distance about 400 light-years), known from classical times as the *Seven Sisters*. However, normal eyes can make out only six stars in the group, the brightest being Eta (η), Alcyone, magnitude 2.9. The Pleiades are too scattered to be an effective telescopic object; powerful binoculars, or a small hand telescope, will give the best view. (NE/B)

The Hyades This star group is scattered over a very large area to the west of Aldebaran. It is much more distant than the brilliant red star itself, which simply happens to lie in the same direction as seen from the Earth. The best view will be obtained with binoculars. (NE/B)

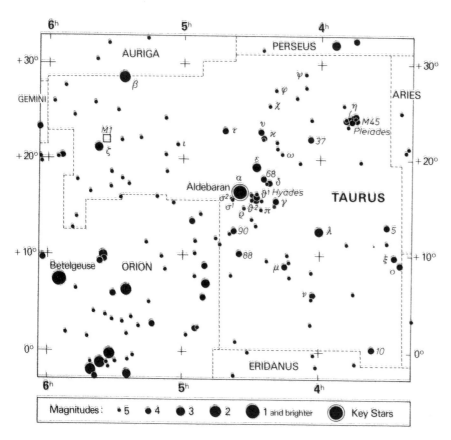

Magnitudes: •5 •4 ●3 ●2 ●1 and brighter ◎ Key Stars

For All Observers:
Cetus (The Whale)

Cetus is an extensive and faintly-marked constellation, its brightest star being Beta (β), Diphda, of magnitude 2.0; Alpha (α), Menkar, is at the opposite end of the group, and is half a magnitude fainter. Diphda is quite conspicuous, since it lies in a barren area of sky far from the Milky Way. The adjoining part of Pisces is also included in this area.

Double star

Gamma (γ) 3.7, 6.2; 3.0″; 295. The brighter star is yellowish. This is a difficult pair, and you should be pleased if you can detect the companion with a 60-mm telescope. (T)

Variable star

Omicron (o) The famous long-period variable Mira, which is mentioned on page 112. It is well worth hunting with binoculars around the time of maximum, when it can rise to anything from magnitude 2 to magnitude 5. With a period of about 330 days, maximum occurs approximately one month earlier each year, and in 1986 and 1987 this is expected to be in March and February respectively, when the star will be in the western sky during the early part of the night. (NE/B)

Galaxy

M77 This is quite a challenge. Some 50 million light-years away, this appears only as a faint speck of light; but it is extraordinary to think that the light making this dim impression on your retina began its journey back in the Eocene period, when the first mammals were developing on the Earth's surface! (T)

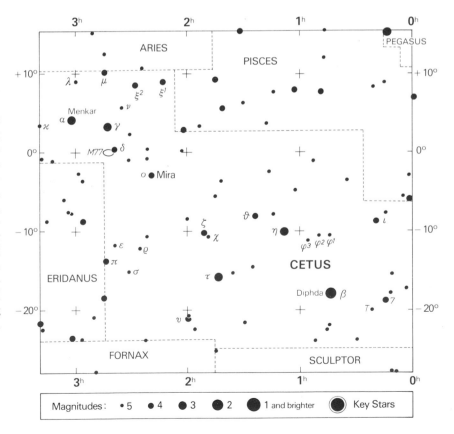

For Southern Observers: Dorado (The Swordfish) and Volans (The Flying Fish)

Neither of these two constellations is conspicuous, but they contain some interesting telescopic objects. Most of Dorado, and all of Volans, is circumpolar from this latitude.

Double stars
(Volans)
Gamma (γ) 3.9, 5.8; 14″; 299. The primary star is yellow. (T)
Epsilon (ε) 4.3, 8.0; 6.1″; 22. The primary star is white. A delicate pair, but not difficult. (T)
Zeta (ζ) 3.9, 9.0; 17″; 116. The primary star is yellowish. It will require considerable application to see the companion with a 60-mm refractor, so the star forms an interesting challenge. (T)

Variable star
(Dorado)
Beta (β) This Cepheid variable (see p. 111) ranges from magnitude 3.4 to 4.1 in the rather long period of 9 days 20 hours. To estimate its brightness, compare it with Alpha (α) Doradus (magnitude 3.3) and Beta (β) Pictoris (magnitude 3.8). (NE/B)

Galaxy
The *Large Magellanic Cloud* or *Nubecula Major* lies in Dorado and Mensa, looking like a detached fragment of the Milky Way, about 7° across. Some of the condensations and nebulae (particularly the amazing NGC 2070, surrounding the star 30 Doradus) can be seen with the naked eye. Telescopically, it is a Pandora's box of clusters and nebulosities, though most require a large aperture for a satisfactory view. This irregular galaxy, a satellite of our own, lies about 180,000 light-years away. (NE/B)

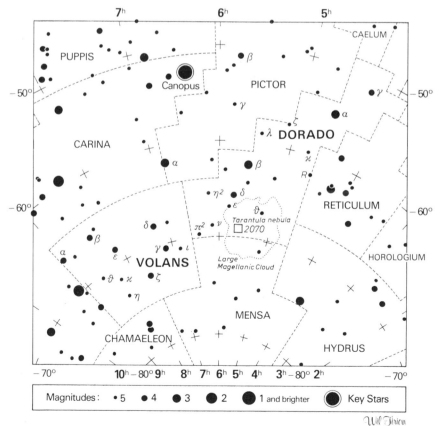

Magnitudes: •5 •4 ●3 ●2 ●1 and brighter ◉ Key Stars

Wil Tirion

7

Astronomical Photography

You can take some sort of celestial photograph with almost any camera. Wait for a clear moonless night, lay the camera on a firm support, point it to the zenith, and leave the shutter open for about twenty seconds. No matter what kind of film you have in it, some stars will be recorded. If the film is 'slow', there may not be many; but a 'fast' film and a wide-aperture lens will easily record all the stars visible, including with this brief exposure not only those seen with the naked eye, but also those visible with moderate binoculars.

So, at this level, photography is quite easy. It is satisfying, too, for a colour slide of any region of the Milky Way can be a glorious sight. The difference between reddish, yellow and white stars can readily be seen, although the fainter white stars usually appear blue on photographs – this does not harm the pictorial appeal, however. But when you look into the subject a bit more closely, you will find that there are a number of aspects to astronomical photography. Taking a picture of a star field, and taking a picture of the Moon, pose entirely different problems. This chapter gives some guidance on the available photographic processes, and then takes a look at how to photograph stars, which requires only the simplest equipment.

Something about Photography

An unexposed photographic film contains silver compounds held in a thin layer known as the *emulsion*. This emulsion is coated on to a transparent plastic backing. Most serious photographers use 35-mm film – in other words, the strip of film which is wound through the camera is 35 mm wide. Each photograph on a film of this size measures 24 mm high and 36 mm wide. Smaller, cheap cameras which take film cartridges use narrower film, also in a roll. Some cameras use cut film, which is made in several different formats, but cut film is expensive and is normally used by professional photographers. Glass plates coated with emulsion were once popular, but nowadays these are made only for special purposes, since they are extremely costly.

After the film has been exposed in the camera, it has to be 'developed', when it is passed through a solution which has a different action on those parts of the emulsion that have been exposed to light, compared with its effect upon those that have not. If the film is a black-and-white type, the result is a *negative*, in which the bright parts of the subject appear dark, and vice versa. The negative is then used to make a positive print on sensitive paper.

In the case of a colour film, there are two possibilities. The film may be turned directly into a *transparency*, which is viewed by holding it up to the light, or projecting it on to a screen. Alternatively, it may be developed as a negative, with the colours apparently all wrong, and then printed on to paper as a positive print.

It is impossible to say which is the best process for astronomical work. Colour slides are very attractive, and the brilliance of a projected picture is much greater than that of a paper print. On the other hand, the negative/positive process, which produces a paper print, is more flexible, since you can enlarge just a small portion of the original negative and make it whatever size you like.

If you are a home photographer, negative/positive work has special attractions. You are in charge of the whole process, and can almost certainly produce a better result than could a wholesale processing firm. If you already have the equipment, black-and-white processing works out at about the same order of cost as colour transparencies, assuming that you do not make very big enlargements. It will probably be best to start with colour transparencies, since, if they work at all, they work very well!

Star Photography with just an Ordinary Camera

Lenses and Films

When taking ordinary photographs, you will probably know that it is most important to get the correct *exposure*. If the image cast on the emulsion is too bright, the resultant picture will be glaring white, and details will be submerged. If the image is too dim, the photograph will be very dark.

So you adjust the amount of light reaching the film by varying the diameter, or *aperture*, of the lens, and by altering the shutter speed, or exposure. The first alters the amount of light passing through the lens, and the second decides for what length of time this light will be allowed to shine on to the film. Various combinations of the two are possible – a lot of light for a short time, or less light for a longer time. The result should be the same. (Take the analogy of filling a glass from a tap: if the water is only trickling, you will have to wait longer than if the tap is full on.)

Aperture is normally rated not in millimetres but in 'f/number'. This is the ratio of the focal length of the lens to its aperture. A typical camera lens may have a maximum aperture of f/2, which means that if its focal length is 50 mm (common with standard 35-mm cameras), its effective aperture is 25 mm. There will be a rotatable ring on the lens barrel which allows this aperture to be reduced, with a minimum opening of, perhaps, f/16, where its aperture will be only 3 mm. If you look inside the lens you will see an adjustable diaphragm to serve this purpose.

The reason for rating a lens aperture in this way is that the brightness of an ordinary terrestrial view depends upon focal ratio, not upon linear aperture. Any f/2 lens, regardless of its focal length, will give an image of a landscape, or of a person, of similar brightness. But this is not true of star photography. Stars are, effectively, points of light, and their brightness depends very strongly upon the linear aperture of the lens and only in a small way upon the focal length. Since a lens of, for example, 60 mm in aperture has four times the area of a 30-mm aperture lens, it will focus four times as much light into a star image, and a similar photograph may be taken in one quarter of the exposure time.

However, you do not need a very large lens in order to take excellent star photographs. A standard 35-mm camera, which will probably have a lens of focal length 50 mm, can take really pleasing pictures of the night sky, and the lens need not even be used at full aperture.

It makes sense to use a 'fast' or highly sensitive film. Slower films have their uses, but not for star photography, since the aim is to use the minimum necessary exposure. If you are taking colour slides, Ektachrome 400 or the more recent 3M Colour Slide 1000 film is recommended. For black-and-white work, load your camera with Tri-X or HP5 film, both of which are of similar speed.

Exposure

An important difference between star photography and any other kind of photography is that there really isn't such a thing as the 'correct' exposure. This is because you are photographing both bright and faint stars at the same time. The longer the exposure, the more intense the star images become, and the fainter the limiting magnitude of the photograph. To take a realistic example, suppose you point a 35-mm camera at the night sky, with an f/2 lens of 50 mm focus, using fast film. Then, a one-second exposure may capture all the stars down to the 6th magnitude. If this is so, then the following table shows *theoretically* how faint a star image you should be able to photograph with longer exposures:

Exposure (seconds)	Limiting magnitude
1	6th
2.5	7th
7	8th
16	9th
40	10th
100	11th

The logic of this table is straightforward enough: stars of each successive magnitude division are $2\frac{1}{2}$ times as bright or faint as stars in the adjacent category, and therefore the exposure required to record them is in a corresponding ratio.

I have said that this table is theoretical. This is because it is assumed that the camera is operating in truly dark conditions. In practice, this never happens. In a town site, a f/2 lens used with fast film will begin to 'fog' in a few seconds. This means that the sky background begins to brighten, and as it brightens so the faintest stars are lost. Even from a dark country site, there is enough residual sky light to produce fogging after a very few minutes' exposure. So you cannot simply go on increasing the exposure without limit, hoping to record fainter and fainter stars! Even so, a photograph showing stars down to the 8th or 9th magnitude is quite impressive as a beginning, and you will certainly be delighted with your first efforts if they show sharp stars as faint as this. Therefore, aim initially at exposures of a few seconds to half a minute or so.

There is another excellent reason for keeping the exposure short: the Earth is turning, and carries your camera with it in a steady arc amounting to one degree of sky every four minutes. Therefore, if the exposure is prolonged the stars will seem to have trailed. Trails may be interesting as a proof that the Earth is spinning, but they are extremely irritating, since the stars begin to overlap each other and the patterns are lost.

Mounting and Guiding the Camera

How should the camera be mounted for a short exposure? An ordinary tripod is only suitable if (a) it is extremely rigid, and (b) it has a fitting at the top which permits the camera to point to a high altitude. Most camera fittings do not, although you may be able to photograph near the zenith by shortening two of the legs so that the tripod lies almost flat. Alternatively, you may be able to improvise something from wood, perhaps holding the camera with stout elastic bands. Necessity is the mother of invention!

Don't forget to fit the camera with some sort of dew-cap. On a damp evening, the lens will quickly become mottled with condensation, and the photographs will be ruined, even though you may not be aware of anything wrong. The camera may be equipped with a lens hood, but this is unlikely to be long enough for the purpose. You can easily extend it with a short ring of black paper – but test through the viewfinder beforehand, to ensure that the dew-cap is not vignetting the field of view. This can be checked by pointing the camera at the daylight sky and noticing if the corners of the field of view are dimmed when you look through the viewfinder.*

*This assumes that you are using a single-lens reflex (SLR) camera. If the camera has a separate viewfinder, a useful rule of thumb is that the projection in front of the lens should not be more than the diameter of the lens mount.

When the camera is loaded with film, and the lens set at maximum aperture (and focused on infinity!), all you need to do is select the sky area to be photographed. You may find it hard to work out what is being included in the view, since only the brightest stars can be seen through a viewfinder anyway, and also, when the camera is pointing high in the sky, you may find it impossible to get your eye anywhere near the viewfinder. Initially, therefore, you may have to trust to luck. Later, you will find it necessary to improvise a finder of some kind, perhaps using a white wire rectangle and a peep-sight (see Fig. 30). It may be useful to remember that the area of sky included on a 35-mm frame, using a 50-mm focal length lens, is about 40° by 28°, or the width of 2 by $1\frac{1}{2}$ outstretched hands.

To give an exposure of a few seconds' duration, you are going to need to hold the shutter open manually. Older-type cameras had a 'T' setting on their shutters, which meant that you could click and release the trigger and the shutter stayed open; a second click and release closed it. This is rarely, if ever, to be found on modern cameras, and the best to hope for is a 'B' setting, which means that the shutter remains open for as long as the trigger is depressed. Use this setting with a locking cable release to hold the shutter open, but *don't begin and end the exposure in this way*. Even the gentlest shutter mechanism can give a fixed camera a jolt, and the typical single-lens reflex 'slam' can vibrate it seriously. The technique is to cover the lens with a black card; open the shutter; and then gently move the card clear of the dew-cap before whisking it aside. This gives time for the shutter vibrations to die down.

If you want a useful exposure of more than about twenty seconds' duration, you will have to guide the camera to follow the stars; for even after this short duration, a star image will show a noticeable elongation. So use your equatorially-mounted telescope, if you have one, to act as a guiding platform. A motor-driven instrument can (in theory at least!) be left to its own devices, but, if the telescope has no drive, you will have to observe the object you are photographing through an eyepiece and keep it fixed in the field of view by turning the slow-motion on the polar axis.

This is not particularly easy, since a star somewhere near the centre of the field of view can drift a small amount without its shift being very noticeable. One technique is to have fine cross-hairs fixed in the eyepiece, and to keep the star firmly behind the intersection. But this involves the fiddly business of making the cross-hairs, and an alternative method, which is certainly good enough for taking photographs with a miniature camera, is to defocus a bright star, so that it looks about the size of Jupiter (or the lunar crater Copernicus) seen with a high magnification, and set it at the edge of the field of view, so that the edge of the star disc just touches the sharp field edge. By keeping the two in contact, you are holding both telescope and camera steady against the Earth's spin.

You may have to devise a way of fixing the camera to the telescope. It needn't necessarily be fitted to the tube; a better arrangement, if you have a small refractor, may be to fit it somewhere on the counterweight, which rotates around the polar axis just as the tube itself does. This means that you will not put the telescope tube out of balance by attaching a heavy weight near its upper end (the camera would have to be near the object-glass end, or else the tube would appear in the photograph!).

More Thoughts about Star Photography

You will probably be delighted at your first efforts; but as you become more experienced and critical you may want to be more systematic and improve your technique. Most amateurs seem to photograph nothing but the Milky Way, which is understandable enough, if you simply wish to include as many stars as possible in a given photograph, but means that huge areas of the sky are rarely portrayed; and attractive groups and combinations of stars can be found everywhere, even in quite barren fields. It is worth considering Lynx or Reticulum as well as Aquila and Sagittarius!

There are also other photographic possibilities, such as recording the movements of the planets; you can photograph Uranus and possibly Neptune as well as the naked-eye planets. Nightly photographs of Mars around the time of opposition will show its retrograde drift as the Earth sweeps past on its inside track, and slower-moving Jupiter and Saturn will reveal movement from week to week. The two outer giants are particularly worth tackling, since very few people have ever observed Uranus and Neptune in this way, and will be intrigued to see their shift against the starry background. A more ambitious project would be to photograph the complete movement of a planet in the course of an apparition, at, say, weekly intervals.

It is also worth experimenting with your technique. Probably, you will have begun your stellar (or planetary!) photography with the camera lens at full aperture. Why not, when this means that the maximum amount of starlight is being collected at a time? It may seem strange to consider reducing the aperture, but there are at least two good reasons for experimenting. Firstly, many lenses give their best definition when stopped down slightly, say to f/4 or f/5. The result should be sharper star images, particularly near the edge of the photograph. Secondly, the degree to which background sky brightness affects the film is enormously affected by aperture ratio. Suppose that a lens used at f/2 begins to produce objectionable sky fog after a thirty-second exposure. If you stop it down to f/4, the same degree of fogging will not be reached for several minutes, whereas it will take only two minutes to photograph stars of the same magnitude as previously. The result will be a much more contrasty and

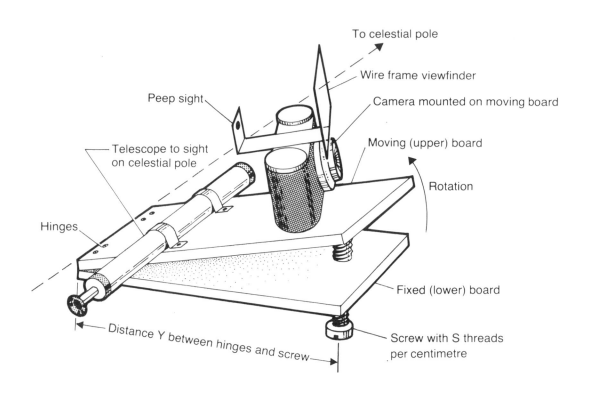

To celestial pole

Wire frame viewfinder

Camera mounted on moving board

Peep sight

Moving (upper) board

Telescope to sight
on celestial pole

Rotation

Hinges

Fixed (lower) board

Distance Y between hinges and screw

Screw with S threads
per centimetre

satisfactory photograph. Also, you could prolong the exposure to record still fainter stars before the dreaded sky fog begins to appear, with the surprising result that a *smaller* lens aperture can photograph *fainter* stars!*

But you will almost certainly need to guide the camera if you are going to reduce the lens aperture, as you will need to lengthen the period of exposure. With very short exposures (twenty seconds, say), all you will do is record *fewer* stars. If you have an equatorial telescope, as we have seen, then all is well, and you can attempt exposures of up to ten minutes or so at f/5 with every chance of results superior to those obtained in, say, two minutes at f/2. Even if you do not possess an equatorial instrument, provided you have a few simple tools, or a friend to wield them for you, a most satisfactory camera mounting can be made quite easily from bits of scrap.

A very simple Equatorial Mounting

This extremely neat and effective device is known as the *Scotch mount*, and was devised by G. Y. Haig of Scotland. Figure 30 is almost self-explanatory.

*This apparent paradox is due to the difference between a point image (that of a star) and an extended image (such as that of an area of sky). When the aperture is reduced, the extended image dims down much more noticeably than does the point image.

30 The principle of the Scotch mount, a very simple device for driving a camera with equatorial motion. If the line of the hinges points towards the celestial pole, turning the screw at the correct rate will make the camera, attached to the moving board, follow the stars.

The Scotch mount consists of two rectangular wooden boards, hinged together at one end. The line passing through the two hinges defines the polar axis, and one board is secured rigidly to a base so that the 'polar axis' can be aligned on the celestial pole. The camera is mounted on the other (moveable) board. A long bolt, or a piece of screwed rod with a knob or handle at the lower end, passes through the fixed board, so that its upper end bears against the underside of the moving board. The bearing end should be rounded as much as possible; a small steel ball cemented to the tip of the rod is even better. When this rod or screw is turned, it imparts a slow rotary motion around the hinges to the upper board, and the camera is carried with it. By choosing the correct rate of turn, an effective one-revolution-per-day is achieved, and the camera is moving equatorially!

The details of the Scotch mount can mostly be determined by the maker. It can be whatever shape and size seems suitable, although the rate at which the drive turns the adjustable board must be correct. However, it is not difficult to calculate this. If the distance between adjacent threads on the screw (the pitch) is S, and it is intended to turn this screw at a rate of T times per minute, then the length Y, which is the distance from the line of the hinges to the point where the screw drive operates on the moveable board, is given by the formula:

$$Y = 228.5 \times S \times T.$$

For example, if the thread has a pitch of 1 mm, and it is decided to turn the screw at a rate of one revolution per minute, the value of Y is 22.85 cm, which is quite a convenient length. The screw can be turned by an electric motor, suitably geared down (a 1 rpm synchronous motor is fairly easy to obtain), but hand-power is just as good, and more portable! The most straightforward way of achieving an accurate hand drive is to arrange for a 1 rpm rate, and to secure a pointer to the screw which is then wound round so that it corresponds to the rotation of the luminous (or faintly-illuminated) hand of a watch or clock. If you plan for this method, ensure that the screw has to rotate clockwise in order to turn the board in the correct direction! When following by hand in this way, it is also essential to wedge and weight the mounting securely, so that it does not quiver as the screw is turned.

The camera can be fitted to the moving board by adapting a ball-and-socket head. As a first approximation, the mounting can be aligned by sighting up the hinges to where the celestial pole is judged to be, but this rather hit-and-miss method can be greatly improved by adding a small sighting telescope to the moving board. This must be aligned with the hinges, with cross-hairs marking the field centre. To check the parallelism between the sighting telescope and the hinges, set up the mounting, bring a bright star or planet to the centre of the field of view, and then swing the moveable board around the hinges. If the adjustment is correct, the object will remain in the centre of the field while this is done. A small telescope certainly makes the setting up of the mounting more certain, for it is annoying to spend a clear night photographing star fields, only to find that misalignment of the polar axis has caused all the stars to trail slightly!

Photography through a Telescope

Photographs taken using an ordinary camera lens show a relatively large area of sky on a very small scale. If the focal length of the lens is F, then one degree of sky (two Sun- or Moon-widths) is represented by a distance on the film equal to $F/57.3$. Therefore, if you are after a solar or lunar image 10 mm across, you need a focal length of about 115 cm. Even this size of image, however, will not show a great deal of detail, although it will certainly be far superior to the naked-eye view. Since the frame on a 35-mm film measures about 36 mm by 24 mm, a 20-mm diameter solar or lunar image could be accommodated comfortably, if the telescope has the necessary focal length of 230 cm or so. However, this would involve a long and awkward tube.

Fortunately, a long focal length can be simulated by enlarging the image formed by a lens or mirror of much shorter focal length, using an eyepiece. The technique is much the same as that involved in projecting an enlarged image of the Sun, as described on page 41, except that in the simple method

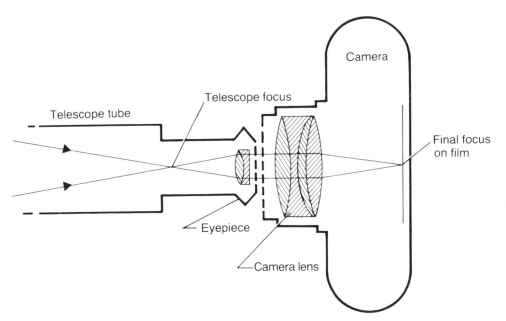

31 How to take a photograph through a telescope. Ensure that the camera is focused on infinity, and that the eyepiece gives a sharp image when the telescope is used in the normal way, with the eye relaxed for distance viewing. When the camera is brought up to the eyepiece, a sharp image should automatically be focused on the film.

to be described here the light goes through the camera lens as well. The arrangement is shown in Figure 31. If the eyepiece is carefully adjusted so that the image is sharp when your eye is relaxed and focused on 'infinity', then the image on the film will also be sharp if the camera is substituted for the eye and its lens set on 'infinity' focus also.

If the camera is of the single-lens reflex (SLR) type, it may be possible to focus the image by looking through the viewfinder, so as to compensate for any errors in your eye's 'infinity' setting. Alternatively, and preferably, take a small hand telescope (or look through one half of a pair of binoculars), obtain as sharp an image as possible of some remote object, and then look through the astronomical telescope, with its eyepiece in place, via the hand telescope. Now focus the eyepiece of the astronomical telescope until the image is as sharp as possible. The view will not be very appealing, since the magnification of the main telescope is multiplied by that of the second instrument, and quite possibly may mean that you are observing with a power of $\times 1000$ or so! But it is a sensitive way of ensuring that the telescope is accurately focused to give a sharp image on the film.

Unless 'snapshots' of the Full Moon or the Sun are being taken, in which case the camera can be held against the eyepiece while the exposure is made, it will be necessary to make or buy some sort of bracket to hold the camera rigidly to the telescope during the exposure.

The amplification given by this eyepiece/camera arrangement is equal to the focal length of the camera lens divided by that of the eyepiece. If you do not know the focal length of the eyepiece, the diameter of the image of the Sun or Moon can be calculated from the following formula:

$$\text{Image diameter} = \text{Magnification} \times F/115$$

where F is the focal length of the camera lens. For example, if the focal length is 50 mm, and it is used in conjunction with a $\times 40$ eyepiece on the main telescope, the solar or lunar image will be about 18 mm across – a very suitable diameter for a 35-mm transparency, since it will almost fill the screen when projected.

Solar photography can be attempted only if a dense filter is available. Special aluminium-coated mylar can be bought from astronomical suppliers, and a piece can be snipped from the sheet and placed loosely over the front aperture of the telescope.* Densely-fogged photographic film is an alternative, but it colours the Sun a deep yellow – and remember that such a filter is certainly *not* safe for visual use. The correct exposure will have to be determined by experiment.

---

*Aluminized mylar is recommended by many people for visual use. As I have already pointed out, there are many different makes of filter available, and you should seek expert advice before using one.

On the whole, the Moon is by far the best candidate for first efforts at astrophotography through a telescope, whether the instrument is fixed or equatorially mounted. Using fast colour film, such as Ektachrome 400, and an aperture of 60 mm, the following exposures (seconds) should be about right:

	Image diameter (mm)			
	5	10	15	20
Full Moon	$\frac{1}{8}$	$\frac{1}{2}$	1	$1\frac{1}{2}$
Quarter Moon	$\frac{1}{2}$	$1\frac{1}{2}$	3	-
Crescent	$1\frac{1}{2}$	3	-	-

If you want to find the exposure for the same image diameter but using a telescope of different aperture (A), multiply the exposures given here by $(60/A)^2$.

Unless a driven equatorial mounting is available, the usable exposure is limited by the speed of the Earth's rotation, which makes the lunar surface appear to drift across the view at a rate of about 26 km per second. Even a one-second exposure, therefore, will blur the image of a moderate crater. However, as the table shows, it is possible to give a much shorter exposure than this, if the Moon is past the quarter stage and the image is not too large. It is also possible to record the crescent phase of Venus with a 'snapshot' exposure, since the cloudy surface of this planet actually appears brighter than that of the Moon.

More ambitious projects, such as photographing clusters and nebulae, or detail on the discs of the planets, require a driven equatorial telescope of at least 150-mm aperture, and take us beyond the scope of this book. But the most successful amateur astrophotographers started their work in the modest kind of way described here, and then proceeded, in the best tradition, to develop their skills and techniques in the ways that brought the best results.

8

Location
by the Stars

A journey across the surface of the Earth will be mirrored by a corresponding motion of the celestial sphere. A journey towards one of the terrestrial poles causes the corresponding celestial pole to rise higher in the sky; head towards the equator, and the pole sinks while the celestial equator elevates itself. Wherever you stand on the Earth's surface, your latitude corresponds to the celestial latitude, or declination, passing through the zenith.

Similarly, a displacement in longitude on the Earth's surface affects the longitude (right ascension) of the celestial sphere which happens to be on the meridian. However, this change must be disentangled from the rotational effect of the Earth, which causes the right ascension on the meridian to change continuously. In order to sort this problem out, you need to have a clock; the story of developments in the art of navigation is very largely one of more sophisticated time-determination. A clock tells you what the right ascension on the meridian ought to be, for your particular terrestrial longitude; if you work with Universal Time (UT), the standard meridian of Greenwich, England, is being used. By noting the actual right ascension that is on your own meridian, and comparing it with what a Greenwich observer would see at the same instance, your distance in degrees of longitude from the Greenwich meridian (longitude 0°) can be determined.

In practice, most people know their own latitude and longitude well enough, although it may still be interesting to try to determine it! Direction-finding, however, could be very important in a number of circumstances, and there are several simple ways in which the celestial bodies can be called upon to help establish the cardinal points on a clear or partly-clear night.

Simple Direction-finding

The celestial poles form the two immoveable direction guides of the sky. The Pole Star, Polaris, lies within 50′ arc of the true pole, and a sighting of this star, which is visible from virtually every place in the northern hemisphere, immediately allows the other cardinal points to be determined. Face Polaris, and the east is on your right hand, south is behind you, and west is to your left. Even if the whole sky is not clear, a glimpse of the two Pointers in the Great Bear (Alpha (α) and Beta (β) Ursae Majoris), or the faint string of stars in the Little Bear (Ursa Minor) that lead to the pole, will be sufficient to identify it (see Fig. 9).

The southern celestial hemisphere lacks a bright pole star, and the position of the pole can be derived only by using nearby marker stars. A line from Alpha (α) Triangulum Australis, carried beyond the little triangle of 4th-magnitude stars formed by Beta (β), Gamma (γ), and Delta (δ) Apodis for the same distance, will arrive within about 3° of the celestial pole. From this, it may be possible to identify the 5th-magnitude star Sigma (σ) Octantis, which currently lies within 57′ of the pole – only a little more distant than its much brighter northern counterpart, Polaris. Another 'fix' can be made using the third point of an equilateral triangle formed from Gamma (γ) Chamaelontis and Alpha (α) Apodis. When facing the south pole, east is on your left hand and west is to your right.

But perhaps the polar regions are obscured by cloud, or are invisible for some other reason. In this case, bear in mind that the celestial equator always cuts the horizon at the east and west points, so that any celestial object known to lie near the equator must indicate the eastern or western horizon when it is rising or setting respectively. Conveniently, the sky's most famous triplet of stars, the 'belt' of Orion, lies practically on the equator, and is bright enough to be seen at a very low altitude, if the sky is transparent.

It may also be helpful to remember that when the outline of Orion, as viewed from temperate latitudes, appears to be standing (or hanging) perfectly upright in the sky, it is about 20° away from the meridian (roughly south-southwest, as seen from northern latitudes, or north-northeast as seen from southern latitudes).

Some other orientation guides that may be of use are as follows:

1 If the line joining Alpha and Gamma Pegasi is horizontal, then they are on either side of the meridian;

2 If the line joining Alpha and Beta Pegasi is vertical, then they lie on the meridian;

3 If Castor and Pollux (Alpha and Beta Geminorum) are vertically above or below Procyon (Alpha Canis Minoris), then the latter is on the meridian;

4 If the line joining Beta Canis Majoris and Canopus (Alpha Carinae) is vertical, then they lie on the meridian.

The Sun, Moon, and planets are of less use as direction indicators, since their positions on the celestial sphere are changing all the time, and their paths along the ecliptic can take them well over 20° north or south of the celestial equator. However, the Sun rises and sets due east and west at the equinoxes (around March 21 and September 23); and, since it is more or less opposite the Sun, the Full Moon also rises and sets within a few degrees of east and west at these times. But the Moon's position is not so predictable as that of the Sun, since the lunar orbit is inclined at an angle of 5° to the ecliptic. The irregular wanderings of the planets mean that their coincidence with the celestial equator cannot be predicted very accurately; but if you are a regular observer, you will certainly know whereabouts on the celestial sphere the brighter planets are to be found.

Determining Latitude

The most straightforward way of determining the latitude of a place is to measure the altitude of the celestial pole above the horizon. A device known as an *alidade* can be used for the purpose. It is a sighting stick carrying a protractor, with a plumb-line to indicate the vertical. Once the celestial pole is brought into sight, and the plumb-line has come to rest, the thread is nipped against the protractor and the value of the inclination read off.

Another way of determining latitude is to measure the altitude of any star of known declination as it crosses the meridian. If its declination is D, and its angle above the southern horizon (or the northern horizon, if observations are being made in the southern hemisphere) is a, then

$$\text{Latitude} = 90 - a \pm D \text{ degrees,}$$

with D being added for a northern observer, and subtracted for a southern one. In order to use this method, you have to know the position of the meridian – but only approximately, since the altitude of a celestial object does not change very much for a quarter of an hour or so on either side of meridian passage.

Probably the simplest, and fundamentally the best, way of measuring latitude would be to construct a *zenith tube*, and if you have a turn for the unusual you might consider this project. Quite simply, a zenith tube is a tube pointing at the zenith! It could be a long piece of plastic drainpipe attached to the wall of a house, set accurately vertical by hanging a plumb line down through it, and used as a naked-eye peep-sight. Alternatively, it could be a true telescope, similarly fixed, with cross-hairs in the eyepiece to define the centre of the field of view, and hence to indicate the zenith. Stars passing through the field as the Earth turns could be identified from an atlas, and their declinations either measured from the map or obtained from a catalogue such as Lampkin's. It would be interesting to find out how accurately such zenith passages could be judged!

Determining Longitude

A zenith tube has more uses than simply for determining latitude: it is also a timing device. The meridian passes through the zenith, so that when a star passes through the centre of the tube's field of view, the sidereal time at that instant must correspond to the star's right ascension: it is 'noon' as measured by that particular star.

Greenwich Sidereal Time at 0h UT throughout the Year

Date		Sidereal Time h m	Date		Sidereal Time h m
Jan	1	6 40	July	4	18 45
	9	7 11		12	19 17
	17	7 43		20	19 48
	25	8 14		28	20 19
Feb	2	8 46	Aug	5	20 51
	10	9 17		13	21 23
	18	9 49		21	21 54
	26	10 21		29	22 26
Mar	6	10 52	Sept	6	22 58
	14	11 24		14	23 29
	22	11 55		22	0 01
	30	12 27		30	0 32
Apr	7	12 58	Oct	8	1 04
	15	13 30		16	1 35
	23	14 01		24	2 07
May	1	14 33	Nov	1	2 38
	9	15 04		9	3 10
	17	15 36		17	3 41
	25	16 07		25	4 13
June	2	16 39	Dec	3	4 44
	10	17 11		11	5 16
	18	17 42		19	5 48
	26	18 14		27	6 19

One sidereal day is equivalent to 23 hours 56 minutes of ordinary time, and any value between these 8-day intervals may be calculated by allowing a difference of 3.9 minutes for each complete day. Due to leap-year adjustments, these values may be in error by up to 2 minutes from one year to the next.

Finding the local sidereal time (LST) is essential if you are to determine your latitude, since the amount by which LST differs from Greenwich Sidereal Time (GST) is a measure of your longitude. The accompanying table gives the value of GST for 0 hours UT on different dates throughout the year, and other dates and times can be interpolated from these. Alternatively, and more accurately, they can be measured on a Philips' planisphere. Suppose, then, that an observer notes the star Alpha Centauri (Rigil Kentaurus) to be on the meridian at 0 hours UT on April 1. He then has two helpful pieces of information:

Greenwich Sidereal Time = 12 hours 34 minutes
RA of Alpha Centauri = 14 hours 38 minutes

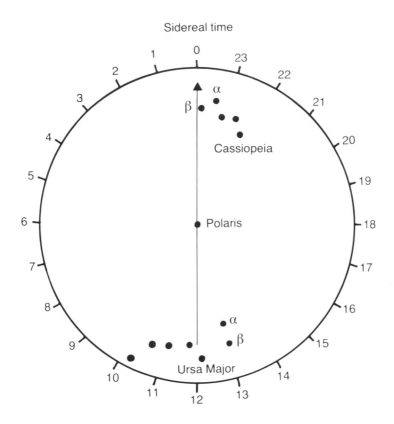

Sidereal time

32 *Sidereal clocks in the sky. In the northern hemisphere the 'pointer' is found by joining stars in Cassiopeia and Ursa Major – when it is vertical, the sidereal time is 0 hours. The pointer rotates anti-clockwise in 24 sidereal hours. In the southern hemisphere, the pointer passes very near Crux and the medium-faint star Beta (β) Hydri, and indicates a sidereal time of 12 hours when it is vertical. This southern 'clock' rotates clockwise once in 24 sidereal hours.*

(The star's RA can either be measured from a map or taken from a catalogue, or, in this instance, found on page 123.) It follows from this observation that the longitude difference between the observer's site and that of Greenwich is equivalent to the amount that the Earth rotates in 2 hours 4 minutes – a fraction over 30 . Since the RA observed on the meridian from the site is *greater* than that of Greenwich, the site must be to the *east* of the prime meridian. Inspection of a globe suggests that a location somewhere along a line of longitude passing through Johannesburg would fit this particular observation.

Determining Sidereal Time

Few people will ever need to derive their longitude in order to find out where they are on the Earth's surface, but it is sometimes useful, for astronomical purposes, to know the Sidereal Time. There is a simple way of determining this fairly accurately, which is by glancing up at the Cosmic Clocks that

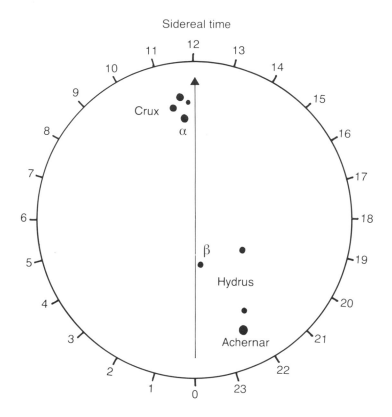

benevolent Nature has arranged for us at each celestial pole. Figure 32 illustrates the northern and southern Clocks.

The northern Clock has, as its pointer, the line joining the Pole Star to Beta (β) Cas (Caiph). Stand facing north, and imagine that Caiph is revolving once in 24 sidereal hours, in an anti-clockwise direction, around an invisible dial centred on the Pole Star. Then 0 or 24 hours is vertically above the pole; 6 hours is to the left; 12 hours is down towards the horizon; and 18 hours is to the right. With care, it is possible to estimate the ST to within half an hour or so.

The southern Clock is not quite so convenient, since the dial has no obvious centre. But, if you can imagine where the pole ought to be, and let Crux sweep clockwise around the imaginary dial, then the ST corresponding to different positions will be as follows (using the faint fourth star in Crux as the pointer): below the pole, 0 or 24 hours; to the left, 6 hours; above, 12 hours; and, to the right, 18 hours.

Some Useful Publications

Star Atlases

A. P. NORTON, *A Star Atlas and Reference Handbook* (Gall & Inglis, Edinburgh: 17th edition, 1978). Definitely the best atlas for the newcomer; it shows about 7000 stars down to magnitude 6.5, as well as numerous deep-sky objects. The many tables and lists are also invaluable. If you can afford only one publication, buy this one!

W. TIRION, *Sky Atlas 2000.0* (Sky Publishing Corp., Cambridge, Mass., 1981). A replacement for the older *Atlas Coeli*, now out of print. It shows about 43,000 stars down to the 8th magnitude, and is an atlas for the more advanced amateur.

Star Catalogues

A. HIRSHFELD and R. W. SINNOTT, *Sky Catalogue 2000.0* (Sky Publishing Corp., Cambridge, Mass., 1981). A companion to the Tirion *Sky Atlas*, this catalogue lists position, magnitude, spectral type, distance, and other information for 45,000 stars down to the 8th magnitude.

R. H. LAMPKIN, *Naked Eye Stars* (Gall & Inglis, Edinburgh, 1972). A very useful catalogue, listing position and magnitude, by constellation, of every star down to magnitude 5.5.

General Handbooks

JAMES MUIRDEN, *The Amateur Astronomer's Handbook* (Harper & Row, New York: 3rd edition, 1983). A comprehensive introduction to practical astronomy for the amateur, with a section on astronomical optical work for those who wish to tackle telescope-making.

IAN RIDPATH and WIL TIRION, *Guide to Stars and Planets* (Collins, London, 1984). A very convenient pocket-sized volume with a detailed description of each constellation, accompanied by a full-page map showing stars down to the 5th magnitude.

Annual Handbooks

The Astronomical Almanac. The standard work, published by the Nautical Almanac Office, Royal Greenwich Observatory, England, and the US Naval Observatory, Washington.

Handbook of the British Astronomical Association (Burlington House, Piccadilly, London).

Observer's Handbook (Sky Publishing Corp., Cambridge, Mass.).

Periodicals

Astronomy (Astromedia Corp., 625 E St Paul Avenue, PO Box 92788, Milwaukee, WI 53202, USA). Monthly.

Popular Astronomy (Quarterly journal of the Junior Astronomical Society, 58 Vaughan Gardens, Ilford IG1 3PD, England).

Sky & Telescope (Sky Publishing Corp., 49 Bay State Road, Cambridge, MA 02238-1290, USA). Monthly.

Appendix

The Time of Local Noon

The most convenient way of determining the direction of the north-south line, as seen from your observing site, is to set up a vertical post (use a plumb line to get it right), and then to set up another post at the tip of the shadow cast by the Sun at local noon. The two posts are then accurately placed on the north-south line, and you can sight along this line to identify some distant marker which is conveniently near the north or south point on the horizon. Once this has been determined, you can always re-establish this important, fundamental direction, as you may need to do if setting up an equatorial telescope by day, when the Pole Star is invisible.

The problem is that the Sun is rarely, if ever, exactly due south (as seen from northern temperate latitudes on the Earth's surface), or due north (as seen from southern temperate latitudes), at 12 o'clock local time. The error may be up to half an hour or so, which corresponds to an error approaching 10° in the derived orientation of the north-south line. There are two reasons for this:

1 Your time system (as shown by the clock on the wall) gives local noon only if your longitude is the standard one (0°, 15°, 30° etc). If it is not, then remember that local noon will be four minutes *early* for every degree you are east of the standard longitude, and four minutes *late* for every degree you are west.

2 The Sun is not always on time! Due to the eccentricity of the Earth's orbit, and the inclination of its equatorial plane to the plane of its orbit, the Sun does not progress at a constant rate around the celestial sphere. The following table shows the error, which you must allow for – it is called the *Equation of Time*, and is given here to the nearest minute.

		Error (m)				*Error* (m)
Jan	1	Slow	3	Mar	2	Slow	12
	6	Slow	6		7	Slow	11
	11	Slow	8		12	Slow	10
	16	Slow	10		17	Slow	8
	21	Slow	11		22	Slow	7
	26	Slow	12		27	Slow	6
	31	Slow	13	Apr	1	Slow	4
Feb	5	Slow	14		6	Slow	3
	10	Slow	14		11	Slow	1
	15	Slow	14		16	-	0
	20	Slow	14		21	Fast	1
	25	Slow	13		26	Fast	2

		Error (m)				*Error* (m)
May	1 Fast	3		Sep	3 -	0
	6 Fast	4			8 Fast	2
	11 Fast	4			13 Fast	4
	16 Fast	4			18 Fast	6
	21 Fast	4			23 Fast	8
	26 Fast	3			28 Fast	9
	31 Fast	2		Oct	3 Fast	11
Jun	5 Fast	2			8 Fast	12
	10 Fast	1			13 Fast	14
	15 -	0			18 Fast	15
	20 Slow	1			23 Fast	16
	25 Slow	2			28 Fast	16
	30 Slow	4		Nov	2 Fast	16
Jul	5 Slow	4			7 Fast	16
	10 Slow	5			12 Fast	16
	15 Slow	6			17 Fast	15
	20 Slow	6			22 Fast	14
	25 Slow	6			27 Fast	12
	30 Slow	6		Dec	2 Fast	11
Aug	4 Slow	6			7 Fast	9
	9 Slow	6			12 Fast	6
	14 Slow	5			17 Fast	4
	19 Slow	4			22 Fast	2
	24 Slow	2			27 Slow	1
	29 Slow	1				

As an example, suppose that you are observing from Exeter, England, on November 2. Exeter is some $3\frac{1}{2}°$ west of the Greenwich meridian. Then, from factor (1) above:

$3\frac{1}{2}°$ is equivalent to fourteen minutes of time. Therefore, local time is fourteen minutes *slow* on Greenwich Time, which is the standard time for the United Kingdom.

From factor (2) above:

The table shows that the Sun will cross the meridian some sixteen minutes *early* on that date.

Combining these two factors we find that the Sun will be due south at two minutes before 12 o'clock noon on that date, so this would be the time to take the shadow measurement.

Now suppose that you are observing from Pittsburgh, USA, on August 4. Pittsburgh is some 5° west of the standard 75° meridian for Eastern Standard Time. Then, from factor (1) above:

5° is equivalent to twenty minutes of time. Therefore, local time is twenty minutes *slow* on Eastern Standard Time.

From factor (2) above:

The table shows that the Sun will cross the meridian six minutes *late* on that date.

Combining these two factors, we find that the Sun will be due south twenty-six minutes after 12 o'clock noon on that date, so this would be the time to take the shadow measurement.

Index

Index of Messier and NGC Objects